Synthesis Lectures on Biomedical Engineering

Series Editor

John Enderle, Storr, USA

This series consists of concise books on advanced and state-of-the-art topics that span the field of biomedical engineering. Each Lecture covers the fundamental principles in a unified manner, develops underlying concepts needed for sequential material, and progresses to more advanced topics and design. The authors selected to write the Lectures are leading experts on the subject who have extensive background in theory, application, and design. The series is designed to meet the demands of the 21st century technology and the rapid advancements in the all-encompassing field of biomedical engineering

Sudip Mukherjee · Vijay Sagar Madamsetty

Nanoparticles in Angiogenesis and Cancer

Springer

Sudip Mukherjee
School of Biomedical Engineering
Indian Institute of Technology (BHU)
Varanasi, Uttar Pradesh, India

Vijay Sagar Madamsetty
Department of Biochemistry
and Molecular Biology
Mayo Clinic College of Medicine
and Science
Jacksonville, FL, USA

ISSN 1930-0328 ISSN 1930-0336 (electronic)
Synthesis Lectures on Biomedical Engineering
ISBN 978-3-031-11283-6 ISBN 978-3-031-11284-3 (eBook)
https://doi.org/10.1007/978-3-031-11284-3

This Springer imprint is published by the registered company Springer Nature Switzerland AG
The registered company address is: Gewerbestrasse 11, 6330 Cham, Switzerland

Preface

Angiogenesis is a physiological process that demonstrates the formation of de-novo blood vessels from preexisting ones. It has a significant contribution in various physiological processes including growth, development, wound healing, etc. Incidentally, angiogenesis is found to play a major role in cancer initiation and progression. Recently, anti-angiogenic based therapeutic agents have shown immense efficacy in cancer treatment. Due to various challenges in the conventional therapies nanomedicine-based therapeutic agents have gained major importance. Consequently, in the last 8–10 years anti-angiogenic nanomaterials were extensively utilized for cancer therapy. This book highlights various anti-angiogenic nanomaterials (including organic, inorganic, liposomes, and protein nanoparticles) and their role in anti-angiogenic cancer therapies. Moreover, the mechanism of anti-angiogenesis is reviewed in detail. Finally, the limitations of these potential therapies and their future perspectives are discussed. We strongly believe that this book will be of immense assistance to a diverse community including students, academicians, researchers, and industrialists.

Varanasi, India
Jacksonville, USA

Sudip Mukherjee
Vijay Sagar Madamsetty

Contents

1 Introduction: Cancer a Global Challenge and Role of Angiogenesis in Cancer

1.1 Introduction

Cancer is a global cause of mortality and morbidity resulting in deaths of millions [1]. Conventional therapeutic approaches have many limitations that restrict the therapeutic options for successful eradication of tumor especially at the later metastatic stages [2–7]. Alternative approaches are constantly being explored that includes immunotherapy, cell therapy, nanomedicine etc. [8]. Like normal tissues, tumors cells constantly require nourishments in form of oxygen and food as a capability to eliminate carbon dioxide and other metabolic excretes. Distinct patterns of tumor-connected neovascularization, achieved by angiogenesis, a natural process that develops new blood vessels from the pre-existing ones, cope with these demands [9]. Many studies validates that the "angiogenic switch" is controlled by compensating components including induction of angiogenesis by the regulation of various signaling proteins such as vascular endothelial growth factor (VEGF) or inhibition of angiogenesis by regulating proteins like thrombospondin-1 etc. [10, 11]. Further the process of angiogenesis gets stimulated in presence of a hypoxic environment or inflammation [12, 13]. Tumor angiogenesis also gets stimulated in presence of other signal factors including fibroblast growth factor (FGF) and transforming growth factor-β, etc. [14]. Among all acceptable ways to cure cancer, anti-angiogenic therapies are an emerging strategy that targets the tumor vasculature and prevents abnormal vessels sprouting in the past few decades [14]. Nevertheless, despite several successful therapeutic clinical uses of anti-angiogenic drugs, hypoxic resistance, unwanted drug toxicity, drug resistance, hindered reaction to radiotherapy, upregulation of different proangiogenic signals, etc., have hampered the prevalence of antiangiogenic therapies [15]. Consequently, nanotechnology presents a striking alternative therapeutic platform, involving smart strategy of delivery vehicles with inimitable physicochemical properties in a targeted manner and long-lasted drug release that holds the potential to

S. Mukherjee and V. S. Madamsetty, *Nanoparticles in Angiogenesis and Cancer*,
Synthesis Lectures on Biomedical Engineering,
https://doi.org/10.1007/978-3-031-11284-3_1

evade the prevailing restrictions [16]. Nanotechnology that involves the use of particles with a size of (1–100 nm) has been extensively utilized in healthcare due to their distinctive structural properties (small size, high surface to volume ratio and high drug loading efficacy, easy surface functionalization, sustainable drug release, multifunctional theranostics properties) [17]. Due to these important advantages various researchers reported diverse biomedical applications of nanomedicine including different diagnostic (biosensing, MRI-imaging, bio-imaging, etc.) and therapeutic (anti-cancer agent, drug delivery, photodynamic therapy, etc.) applications for the treatment of cancer [16, 17].

In the present book, we shall thoroughly discuss the restrictions of current anti-angiogenic cancer therapies, use of nanomedicine as an alternative mode including various organic (liposomes, lipid and polymer nanoparticles, and protein nanoparticles etc.) and inorganic nanoparticles (gold, silver, carbon-based nanoparticles, silica nanoparticles, rare earth nanoparticles, etc.), their safety, limitations, clinical outlook, and future perceptions.

1.2 Cancer, Statistics, Conventional Therapy, Challenges

Cancer and malignancy is the foremost reason of mortality all over the world and is the reason for an immense health and financial liability. As per the World Health Organization (WHO) an estimated 9.6 million people passed away due to various cancers in 2018. In 2021–22 alone, an estimated 1,898,160 cancer incidents were reported and nearly 608,570 cancer linked deaths happened in the United States [1]. The data presented by the National Center for Health Statistics (NCHS) shows a general reduction in cancer occurrence by 27% over the time of last 25 years. Conventional treatment of cancer includes the application of chemotherapy, radiotherapy, surgery either alone or in combinations. However, these conventional therapies possess many limitations toxic side effects, drug resistance, lack of targeting, lack of efficiency, etc. [2–7]. The statistics shown by WHO, and NCHS demonstrates that an early detection and diagnosis of cancer together with improved treatment approaches can significantly increase the chance of cure from cancer [18]. Many alternative approaches have been utilized in recent years including immunotherapy, cell-therapy, checkpoint inhibitor therapy, etc. However, high cost, lack of detail information still hinders successful clinical applications of these newly developed treatment strategies.

As we all know that angiogenesis stimulates the tumor progression and is the main reason for metastasis that increase the mortality rate [19]. Recent alternative therapies were introduced that includes a single or combination approaches of anti-angiogenic drugs/antibodies and anti-lymphangiogenic treatments to inhibit the progression of cancer [15]. However, these therapies with anti-angiogenic drugs also have several limitations and alternative nanomedicine approaches are on the rise as it presents many advantages over the present strategies. Contemplating the achievement of antiangiogenic therapies and improvements created in nanoparticles-based therapeutics towards the inhibition of new blood vessel formations during tumor angiogenesis, the anti-angiogenic nanomedicine can play a massive role to decrease the cancer mortality and morbidity [16, 17, 20].

1.3 Definition of Angiogenesis

Angiogenesis is a physiological procedure that facilitates the formation of de-novo blood vessels from the already existing vasculatures [16, 17, 20]. 'Angio' derives from the Greek word 'angeio' that implies 'blood vessel' whereas, 'genesis' signifies 'production' [21]. The procedure of angiogenesis comprises of numerous vital stages, (a) endothelial cell stimulation in presence of growth factors, (b) endothelial cell proliferation, (c) endothelial cell migration, (d) capillary tube construction and (e) extracellular matrix degradation in presence of various proteolytic enzymes including proteases, peptidases, or proteinases [21]. Biologically, a healthful human body retains the balance of angiogenesis by sustaining the pro- and anti-angiogenic signals. Interrupted equilibrium of angiogenesis can cause serious diseases including generation of tumor, inflammatory disorders, immune disorders, ischemia, and other infections.

1.4 Types of Vessel Formation

Blood vessel formation in healthy body is a naturally occurring process and is responsible for many crucial physiological activities. Blood vessel formation can occur utilizing different steps during angiogenesis namely sprouting angiogenesis, vasculogenesis along with intussusception (Fig. 1.1) [21]. Sprouting angiogenesis, which involves the generation of de-novo capillaries from the already existing ones is the simple form of angiogenesis [22]. During tumor progression, endothelial cells (ECs) are stimulated by certain angiogenic growth factors which in turn trigger the release of different protease enzymes, causing the degradation of the ECM as well as the basement tissue. This requires numerous subsequent stages including (i) formation of tip cells, (ii) invasion of tip cells, (iii) migration of tip cells, (iv) sprouting of tip cells followed by fusion and lumen formation, and (v) stabilization and growth through the recruitment of mural cells followed by formation of extracellular matrix (ECM) [22]. The sprouting angiogenesis is significant during the formation, invasion, and metastasis of tumors. This overall process is firmly regulated by positive and negative regulators, the equilibrium controls the degree of continuing angiogenesis.

The vascular structure originates from two fundamental procedures: angiogenesis and vasculogenesis [24–26]. Vasculogenesis is a physiological process that causes the differentiation of endothelial originator cells or angioblasts into the ECs. Further a vascular network forms in this process of vasculogenesis. Intussusception is a form of splitting angiogenesis where de-novo blood vessels and capillaries are developed from the present ones resulting in the remodeling of blood vessels [27]. The splitting of vessels due to the insertion of tissue pillars cause the division and develops new blood vessels. Additional patterns of angiogenesis and blood vessels formations are reported mainly during tumor angiogenesis including vascular mimicry, vessel co-option, and EC differentiation from

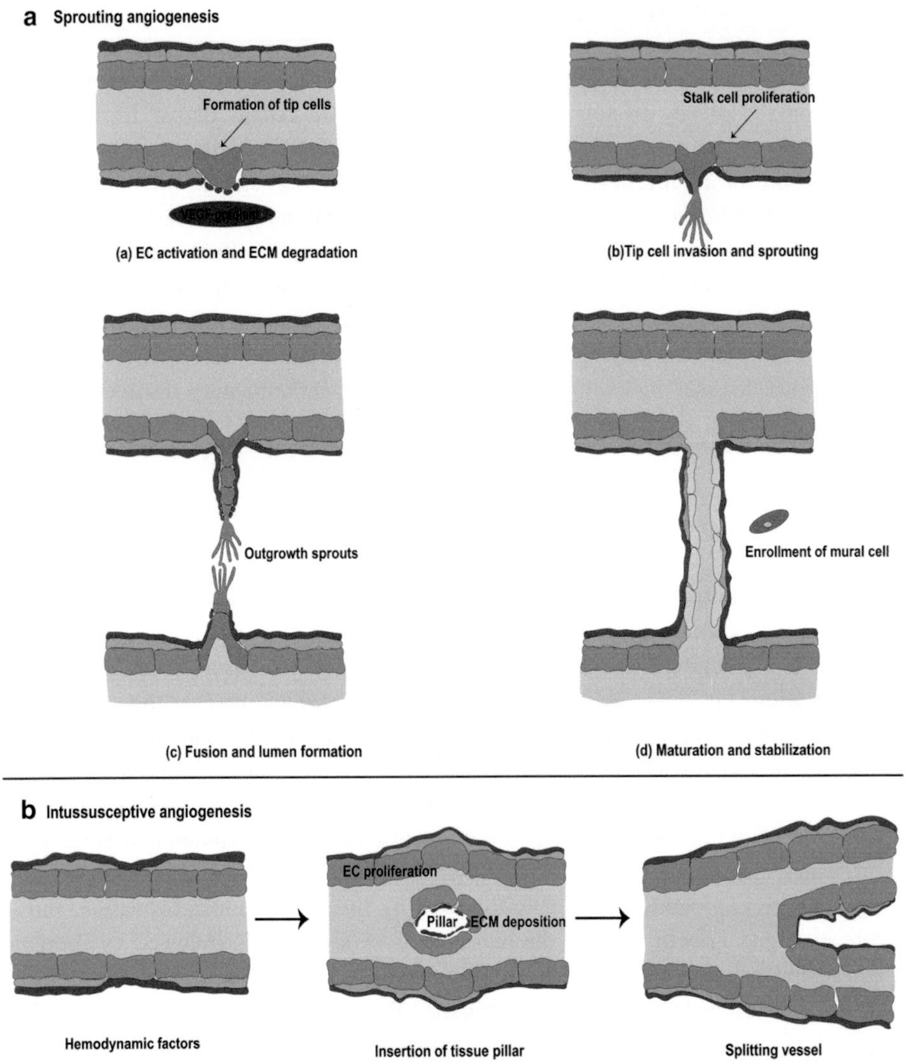

Fig. 1.1 Schematic model of **a** sprouting angiogenesis and **b** intussusceptive angiogenesis. Figure is reproduced with permission from Springer Nature publishers [23]

the tumor stem cells. It is important to discuss about different types of blood vessel formations and angiogenesis before highlighting different use of anti-angiogenic drugs and nanomedicine for the treatment of cancer.

1.5 Angiogenic Stimulators

Endothelial cell proliferation, cell migration, tube formation are the pivotal steps for angiogenesis that is often stimulated by various pro-angiogenic factors present in the body. These can be called a naturally present angiogenic stimulators that includes common growth factors including vascular endothelial growth factor (VEGF), hepatocyte growth factor (HGF), fibroblast growth factor (FGF), insulin like growth factor, granulocyte colony stimulating factor (G-CSF), angiopoietin (Ang1 and Ang2), integrins etc. (Table 1.1). Consequently, angiogenic therapies can play an important role to treat various cancers in human where angiogenesis is influenceable for tumor growth, expansion, and metastasis.

Table 1.1 Angiogenic stimulators and their role

Angiogenic stimulator	Role	References
VEGF	Angiogenesis and lymphangiogenesis inducer	[28]
FGF	Regulates endothelial cell proliferation, migration, and differentiation	[29, 30]
HGF	Angiogenic stimulator, promotes cell growth	[31]
Ang1 and Ang2	Angiogenic regulator, promotes matured vessel formation	[32]
PDGF	Angiogenic stimulator, regulates cellular growth and division	[33]
IGF	Angiogenic stimulator, regulates myogenesis and nerve regeneration	[34]
Endoglin	Stimulates endothelial cell proliferation and ECM production	[35]
Interleukin 8	Stimulates endothelial cell proliferation and MMP production	[36]
Thyroxin	Stimulates primary coronary angiogenesis	[37]
G-CSF	Stimulates endothelial cell proliferation	[38]
Integrins	Promotes cell attachment, stimulates cell migration	[39]
YKL40	Stimulates tumor angiogenesis, has crucial function towards radio-resistance and glioblastoma progression	[40]
HIF1α	Controls tumor angiogenesis and invasion	[41]
TGF-α/β	Inducer of angiogenesis by VEGF-mediated apoptosis. Suppressor of tumor at an early stage but stimulates tumor progression at a later stage	[42]

1.6 Tumor Angiogenesis

Judah Folkman and coworkers in 1971, proposed that tumor progression and metastasis can be associated with the tumor angiogenesis process [10, 11, 43]. It has been seen that a primary tumor has ability to grow upto 1–2 mm^3 gaining necessary source of diffusive nutrients and oxygen. However, for a tumor to grow past this size requires tumor vascularization by the process called tumor angiogenesis. Consequently, the tumor becomes an angiogenic phenotype by the process called "angiogenic switch" that entices blood vessels from the neighboring stroma [44, 45]. Tumor angiogenic switch is regulated by a mixture of pro- and anti-angiogenic factors, a must prerequisite for additional tumor growth. Pro-angiogenic stimulators play crucial role for the tumor-growth while down-regulation of anti-angiogenic inhibitors furthers triggers faster tumor expansion and eventual metastasis. Clearly, angiogenesis is pivotal in regulating tumor growth, expansion, survival and metastasis in human body. Hence, aiming to target the tumor angiogenesis is considered as rational for treating different malignancies that has led to the development of several anti-angiogenic drugs/agents currently either clinically approved or being used in clinical trials.

Hypoxia is an important state that can severely modulate the tumor conditions. Under hypoxic or ischemic conditions, the tumor angiogenesis process gets stimulated where tissue damages have been observed [46]. Tumor cells generally expand quickly with hypoxic in character, which produce vascular endothelial growth factor A (VEGFA). It is important to mention that tumor stromal cells also can generate these pro-angiogenic factors along with tumor cells that triggers tumor growth and expansion. The endothelial cells present inside the tumor microenvironment express VEGF receptor (VEGFR2) that can sense the VEGF gradient using VEGF-VEGFR interactions [47]. As a results de-novo endothelial sprouts forms in the direction of the tumor cells.

Tumor cell generates a complicated cataract of angiogenic signaling pathways that triggers downstream various cellular events in numerous cell types, inducing in the endothelial cells, causing the tumor angiogenesis (Fig. 1.2). The separation of the perivascular cells from the existing blood vessels instigates the proliferation of endothelial cells along with remodeling of existing blood vessels followed by activation of platelets to be enlisted to the basement membranes. Further, tumor-associated macrophages (TAM) produce several angiogenic factors including VEGF and Matrix metalloproteinases (MMPs) causing tumor growth, ECM degradation, invasion and metastasis. Alternatively, endothelial precursor cells activated by the pro-angiogenic factors that leads to the release of proteases causing ECM remodeling and directional sprouting. This in turn promotes the tube formation, vessels branching causing maturation. Additionally, cancer progenitor cells differentiate to endothelial cells, and gets directly involved in the process of angiogenesis [48].

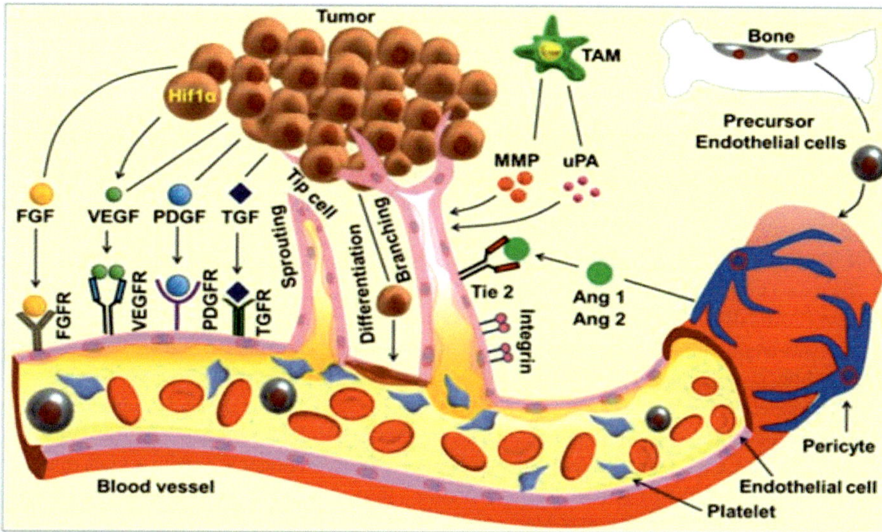

Fig. 1.2 Angiogenic signaling pathway and angiogenesis. Figure is reproduced with permission from MDPI publishers [16]

References

1. R.L. Siegel, K.D. Miller, H.E. Fuchs, A. Jemal, Cancer statistics, 2022, CA Cancer J Clin 72(1) (2022) 7–33.
2. C.L. Arteaga, C.I. Truica, Challenges in the development of anti-epidermal growth factor receptor therapies in breast cancer, Semin Oncol 31(1 Suppl 3) (2004) 3–8.
3. G. Batist, J.H. Wu, A. Spatz, W.H. Miller, Jr., E. Cocolakis, C. Rousseau, Z. Diaz, C. Ferrario, M. Basik, Resistance to cancer treatment: the role of somatic genetic events and the challenges for targeted therapies, Front Pharmacol 2 (2011) 59.
4. G. Chhabra, A. Eggert, N. Puri, Clinical Challenges to Current Molecularly Targeted Therapies in Lung Cancer, Arch Cancer Res 3(3) (2015).
5. J. Rivera-Concepcion, D. Uprety, A.A. Adjei, Challenges in the Use of Targeted Therapies in Non-Small Cell Lung Cancer, Cancer Res Treat 54(2) (2022) 315–329.
6. R.L. Schilsky, J. Allen, J. Benner, E. Sigal, M. McClellan, Commentary: tackling the challenges of developing targeted therapies for cancer, Oncologist 15(5) (2010) 484–7.
7. P. Workman, The opportunities and challenges of personalized genome-based molecular therapies for cancer: targets, technologies, and molecular chaperones, Cancer Chemother Pharmacol 52 Suppl 1 (2003) S45–56.
8. M. Ferrari, Cancer nanotechnology: opportunities and challenges, Nat Rev Cancer 5(3) (2005) 161–71.
9. C.C. Cheng, W.T. Chao, C.C. Liao, J.H. Shih, Y.S. Lai, Y.H. Hsu, Y.H. Liu, The Roles Of Angiogenesis And Cancer Stem Cells In Sorafenib Drug Resistance In Hepatocellular Carcinoma, Onco Targets Ther 12 (2019) 8217–8227.

10. J. Folkman, Angiogenesis in cancer, vascular, rheumatoid and other disease, Nat Med 1(1) (1995) 27–31.
11. J. Folkman, Angiogenesis: an organizing principle for drug discovery?, Nat Rev Drug Discov 6(4) (2007) 273–86.
12. M. Petrillo, F. Patella, F. Pesapane, M.B. Suter, A.M. Ierardi, S.A. Angileri, C. Floridi, M. de Filippo, G. Carrafiello, Hypoxia and tumor angiogenesis in the era of hepatocellular carcinoma transarterial loco-regional treatments, Future Oncol 14(28) (2018) 2957–2967.
13. H.F. Dvorak, Angiogenesis: update 2005, Journal of Thrombosis and Haemostasis 3(8) (2005) 1835–1842.
14. N. Ferrara, A.P. Adamis, Ten years of anti-vascular endothelial growth factor therapy, Nature Reviews Drug Discovery 15(6) (2016) 385–403.
15. A.M.E. Abdalla, L. Xiao, M.W. Ullah, M. Yu, C. Ouyang, G. Yang, Current Challenges of Cancer Anti-angiogenic Therapy and the Promise of Nanotherapeutics, Theranostics 8(2) (2018) 533–548.
16. A. Mukherjee, V.S. Madamsetty, M.K. Paul, S. Mukherjee, Recent Advancements of Nanomedicine towards Antiangiogenic Therapy in Cancer, Int J Mol Sci 21(2) (2020).
17. S. Mukherjee, C.R. Patra, Therapeutic application of anti-angiogenic nanomaterials in cancers, Nanoscale 8(25) (2016) 12444–70.
18. C.P. Wild, The global cancer burden: necessity is the mother of prevention, Nat Rev Cancer 19(3) (2019) 123–124.
19. D.R. Bielenberg, B.R. Zetter, The Contribution of Angiogenesis to the Process of Metastasis, Cancer J 21(4) (2015) 267–73.
20. S.K. Nethi, A.K. Barui, S. Mukherjee, C.R. Patra, Engineered Nanoparticles for Effective Redox Signaling During Angiogenic and Antiangiogenic Therapy, Antioxid Redox Signal 30(5) (2019) 786–809.
21. P. Carmeliet, R.K. Jain, Principles and mechanisms of vessel normalization for cancer and other angiogenic diseases, Nature Reviews Drug Discovery 10(6) (2011) 417–427.
22. F. Hillen, A.W. Griffioen, Tumour vascularization: sprouting angiogenesis and beyond, Cancer Metastasis Rev 26(3–4) (2007) 489–502.
23. M. Ahmadi, J. Rezaie, Tumor cells derived-exosomes as angiogenenic agents: possible therapeutic implications, J Transl Med 18(1) (2020) 249.
24. J.F. Arnold, V. Marmolejo, Visualization of angiogenesis and vasculogenesis in a late tissue radiation injury of the chest wall treated with adjuvant hyperbaric oxygen therapy using fluorescence angiography, Undersea Hyperb Med 46(1) (2019) 69–73.
25. A.M. Dumitrescu, C.F. Costea, C. Furnica, M.D. Turliuc, A.I. Cucu, C.M. Bogdanici, S. Turliuc, G.F. Dumitrescu, A. Sava, Morphological aspects of the vasculogenesis and angiogenesis during prenatal edification of the circle of Willis: a review, Rom J Morphol Embryol 62(3) (2021) 679–687.
26. E. Zarychta, B. Ruszkowska-Ciastek, Cooperation between Angiogenesis, Vasculogenesis, Chemotaxis, and Coagulation in Breast Cancer Metastases Development: Pathophysiological Point of View, Biomedicines 10(2) (2022).
27. V. Djonov, M. Schmid, S.A. Tschanz, P.H. Burri, Intussusceptive angiogenesis: its role in embryonic vascular network formation, Circ Res 86(3) (2000) 286–92.
28. A. Hoeben, B. Landuyt, M.S. Highley, H. Wildiers, A.T. Van Oosterom, E.A. De Bruijn, Vascular endothelial growth factor and angiogenesis, Pharmacol Rev 56(4) (2004) 549–80.
29. M.J. Cross, L. Claesson-Welsh, FGF and VEGF function in angiogenesis: signalling pathways, biological responses and therapeutic inhibition, Trends Pharmacol Sci 22(4) (2001) 201–7.

30. T. Fujii, Y. Yonemitsu, M. Onimaru, M. Inoue, M. Hasegawa, H. Kuwano, K. Sueishi, VEGF function for upregulation of endogenous PlGF expression during FGF-2-mediated therapeutic angiogenesis, Atherosclerosis 200(1) (2008) 51–7.
31. R. Morishita, H. Makino, M. Aoki, N. Hashiya, K. Yamasaki, J. Azuma, Y. Taniyama, Y. Sawa, Y. Kaneda, T. Ogihara, Phase I/IIa clinical trial of therapeutic angiogenesis using hepatocyte growth factor gene transfer to treat critical limb ischemia, Arterioscler Thromb Vasc Biol 31(3) (2011) 713–20.
32. T. Hato, M. Tabata, Y. Oike, The role of angiopoietin-like proteins in angiogenesis and metabolism, Trends Cardiovasc Med 18(1) (2008) 6–14.
33. N. Sato, J.G. Beitz, J. Kato, M. Yamamoto, J.W. Clark, P. Calabresi, A. Raymond, A.R. Frackelton, Jr., Platelet-derived growth factor indirectly stimulates angiogenesis in vitro, Am J Pathol 142(4) (1993) 1119–30.
34. S. Shigematsu, K. Yamauchi, K. Nakajima, S. Iijima, T. Aizawa, K. Hashizume, IGF-1 regulates migration and angiogenesis of human endothelial cells, Endocr J 46 Suppl (1999) S59–62.
35. F. Nassiri, M.D. Cusimano, B.W. Scheithauer, F. Rotondo, A. Fazio, G.M. Yousef, L.V. Syro, K. Kovacs, R.V. Lloyd, Endoglin (CD105): a review of its role in angiogenesis and tumor diagnosis, progression and therapy, Anticancer Res 31(6) (2011) 2283–90.
36. A. Li, S. Dubey, M.L. Varney, B.J. Dave, R.K. Singh, IL-8 directly enhanced endothelial cell survival, proliferation, and matrix metalloproteinases production and regulated angiogenesis, J Immunol 170(6) (2003) 3369–76.
37. R.J. Tomanek, M.K. Doty, A. Sandra, Early coronary angiogenesis in response to thyroxine: growth characteristics and upregulation of basic fibroblast growth factor, Circ Res 82(5) (1998) 587–93.
38. S.T. Lee, K. Chu, K.H. Jung, S.Y. Ko, E.H. Kim, D.I. Sinn, Y.S. Lee, E.H. Lo, M. Kim, J.K. Roh, Granulocyte colony-stimulating factor enhances angiogenesis after focal cerebral ischemia, Brain Res 1058(1-2) (2005) 120–8.
39. C.J. Avraamides, B. Garmy-Susini, J.A. Varner, Integrins in angiogenesis and lymphangiogenesis, Nat Rev Cancer 8(8) (2008) 604–17.
40. R.A. Francescone, S. Scully, M. Faibish, S.L. Taylor, D. Oh, L. Moral, W. Yan, B. Bentley, R. Shao, Role of YKL-40 in the angiogenesis, radioresistance, and progression of glioblastoma, J Biol Chem 286(17) (2011) 15332–43.
41. R. Du, K.V. Lu, C. Petritsch, P. Liu, R. Ganss, E. Passegue, H. Song, S. Vandenberg, R.S. Johnson, Z. Werb, G. Bergers, HIF1alpha induces the recruitment of bone marrow-derived vascular modulatory cells to regulate tumor angiogenesis and invasion, Cancer Cell 13(3) (2008) 206–20.
42. G. Ferrari, B.D. Cook, V. Terushkin, G. Pintucci, P. Mignatti, Transforming growth factor-beta 1 (TGF-beta1) induces angiogenesis through vascular endothelial growth factor (VEGF)-mediated apoptosis, J Cell Physiol 219(2) (2009) 449–58.
43. J. Folkman, Tumor angiogenesis: therapeutic implications, N Engl J Med 285(21) (1971) 1182–6.
44. D. Hanahan, R.A. Weinberg, Hallmarks of cancer: the next generation, Cell 144(5) (2011) 646–74.
45. J. Harper, M.A. Moses, Molecular regulation of tumor angiogenesis: mechanisms and therapeutic implications, EXS (96) (2006) 223–68.
46. M. De Palma, D. Biziato, T.V. Petrova, Microenvironmental regulation of tumour angiogenesis, Nat Rev Cancer 17(8) (2017) 457–474.
47. M. Potente, P. Carmeliet, The Link Between Angiogenesis and Endothelial Metabolism, Annu Rev Physiol 79 (2017) 43–66.
48. L. Yadav, N. Puri, V. Rastogi, P. Satpute, V. Sharma, Tumour Angiogenesis and Angiogenic Inhibitors: A Review, J Clin Diagn Res 9(6) (2015) XE01–XE05.

Current Anti-angiogenic Therapies and Their Limitations

2

2.1 Anti-angiogenic Therapy: Conventional Treatments, and Mechanisms

The process consist of angiogenesis presents the supply of blood to malignant cells promoting cancer development, invasion, and subsequent metastasis [1]. The cancer cells live on high supply of blood and other nutrients. Without constant supply of those tumor development can't progress rapidly [2]. It is clearly recognized that due to release of tumor angiogenesis factor (TAF) from cancer cells, can instigate endothelial cell proliferation and migration to the host blood vessels. Subsequently, the neighboring pre-cancerous cells and tissues turn out to be malignant following procuring these angiogenic capabilities [2]. Additionally, tumor cells produce many other tumor-derived growth factors including VEGF that stimulates endothelial cell proliferation, inflammation, vasculogenesis, vascular invasion and pro-angiogenic switch. Endogenous equilibrium between pro-angiogenic and anti-angiogenic signaling, and associated molecules play a major role in advance of 'angiogenic switch' [3, 4].

To inhibit the pro-angiogenic tumor progression research to design and develop anti-angiogenic drugs have made rapid progress since last decade. The invention to develop drugs that cause anti-angiogenesis (obstruction of blood supply) is useful to inhibit cancer. Importantly, vascular endothelial growth factor (VEGF) is extremely over-expressed in different tumors, when get attached to the vascular endothelial growth factor receptor (VEGFR), triggers the cancer cell proliferation and de novo blood vessels growth [5–7]. The discovery of anti-angiogenic drugs has been mostly focused on the development of VEGF inhibitors that prevents these previously mentioned cascades of events to control rapid tumor progression. Various drugs with the angiogenesis inhibitory properties were identified that had been clinically utilized in the treatment of cancers. Bevacizumab (avastin), an anti-VEGF monoclonal antibody was clinically approved

S. Mukherjee and V. S. Madamsetty, *Nanoparticles in Angiogenesis and Cancer*,
Synthesis Lectures on Biomedical Engineering,
https://doi.org/10.1007/978-3-031-11284-3_2

in 2004 for metastatic colorectal cancer and glioblastoma [8]. Bevacizumab attaches to VEGF to inhibit the stimulation of VEGFR by limiting its binding with VEGFR. Other FDA approved inhibitors include sunitinib, sorafenib, and pazopanib that inhibits VEGFR-2 tyrosine kinase to regulate the tumor angiogenesis. Thalidomide has been found to inhibit VEGF production, while Temsirolimus decreases the production of VEGF. Lenalidomide can inhibits VEGF-induced PI3K-Akt pathway to regulate cancer progression and tumor growth.

'Vascular-disrupting agent' (VDA) are another new class of anti-angiogenic agents that have been recently found to generate severe vascular occlusion that disrupts blood flow in the tumor area [9]. These new drugs including ABT-571, MN-029, ZD6126, AVE8062, combretastatin, and 5,6-dimethylxanthenone- 4-acetic acid have shown remarkable effects in patients and are currently used in various clinical trials [9]. Endogenous anti-angiogenic factors are beneficial for the reticence of cancer growth and progression. Endostatin induces apoptosis in endothelial cells to inhibit their migration, while angiostatin inhibits the functions of VEGF and bFGF. Thalidomide is an anti-angiogenic drug that inhibits angiogenesis by reducing the levels of VEGF, COX-2, bFGF and tumor necrosis factor (TNF-α). For more detailed read, please refer to the excellent review by Clarke et al. that summarized different examples of angiogenic inhibitors including tyrosine kinase inhibitors, monoclonal antibodies, VDA agents etc. and their applications with current clinical status. The application of various angiogenesis inhibitors might be beneficial as encouraging treatment approach for cancer research. A list of various anti-angiogenic drugs, their action target and mechanisms of actions are provided in Table 2.1.

2.2 Limitations of Conventional Anti-angiogenic Therapy

Conventional therapies for cancer have various challenges including drug resistance, drug toxicity, lack of targeting, lack of availability of biomarkers etc. that can create troubles for an effective therapeutic approach [28]. Recently, the anti-angiogenic therapy by impeding the VEGF pathway has developed the extremely significant and commonly recognized method for the therapy of cancer [29]. Though, the medical use of these techniques is yet insufficient since different challenges including harmful consequences, toxicity, developed drug resistance, and lack of availability of suitable biomarkers. However, Bevacizumab (Avastin), the initial (VEGF) directed anti-angiogenesis medication, was clinically granted to treat cancer, only to fail because of developing drug-resistance [30]. This main challenge caused surge in research to understand new pathways including VEGF-independent angiogenesis. Researchers have discovered that VEGF-A can be swapped by other VEGF group gene members including VEGF-C or VEGF-D that binds to VEGFR and activate the connecting cascades [31]. Additionally, VEGF-A may be switched by various

Table 2.1 Anti-angiogenic agents for the treatment of different tumors

Name	Structure	Function	Target cancer	References
Thalidomide	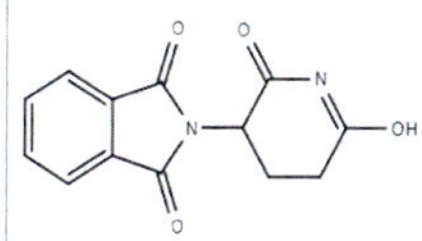	Inhibits VEGF production	Multiple myeloma	[10]
Lenalidomide	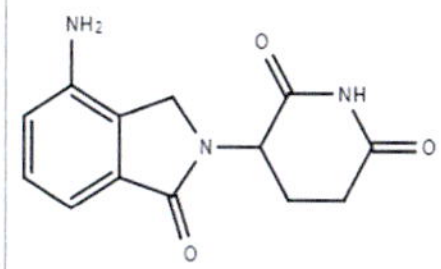	Inhibits VEGF-induced PI3K-Akt pathway	Lymphoma, Multiple myeloma	[11]
Bevacizumab	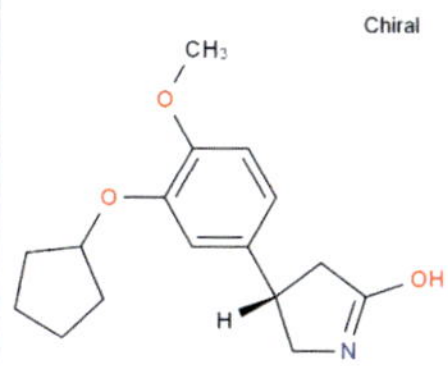	Anti-VEGF monoclonal antibody	Cervical cancer, Colorectal cancer, Glioblastoma, non-small cell lung cancer, Ovarian Cancer, Renal cancer	[12, 13]
Sunitinib	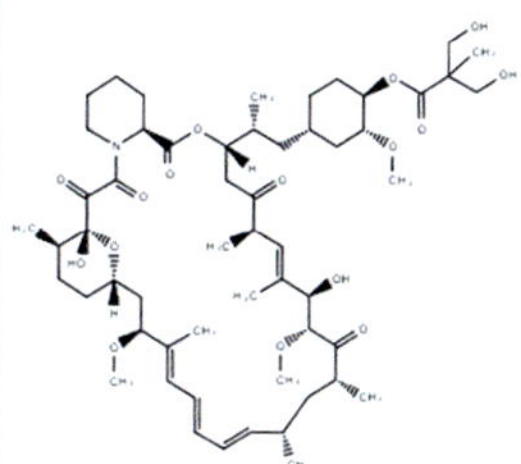	Inhibits production of VEGFR-2 tyrosine kinase	Pancreatic cancer, Gastrointestinal stromal tumor, Renal cell carcinoma	[14]
Temsirolimus	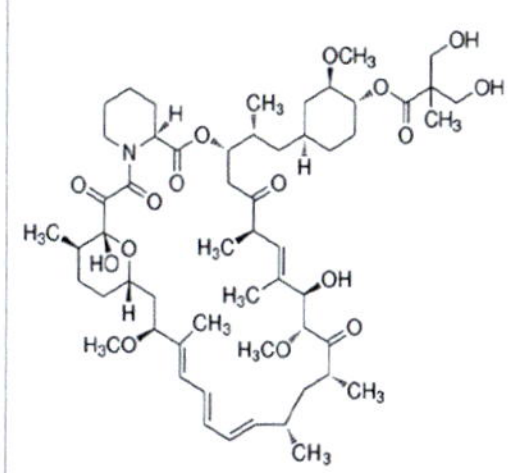	Decreases production of VEGF	Retinoblastoma. Renal cell carcinoma	[15]
Sorafenib	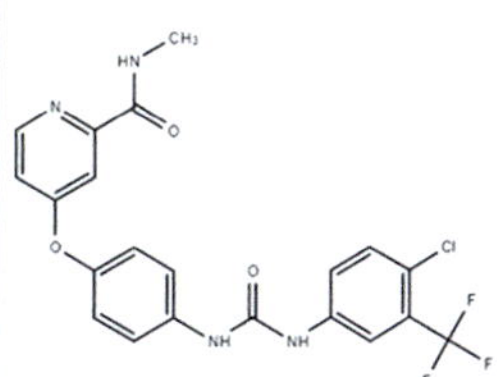	Inhibits production of VEGFR-2 tyrosine kinase	Hepatocellular cancer, Renal cell carcinoma, Thyroid cancer	[16]

(continued)

Table 2.1 (continued)

Name	Structure	Function	Target cancer	References
Lenvatinibmesylate	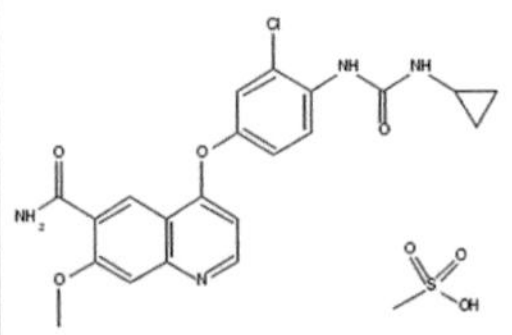	Inhibits tyrosine kinase activity of VEGF1, VEGF2 VEGF3, and (FGFRs)	Hepatocellular carcinoma, Renal cancer	[17]
Pazopanib	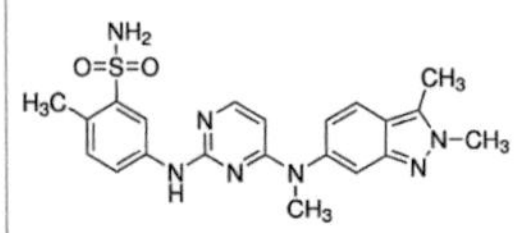	Tyrosine kinase inhibitor	Renal cell carcinoma, Soft tissue sarcoma	[18]
Cabozantinib	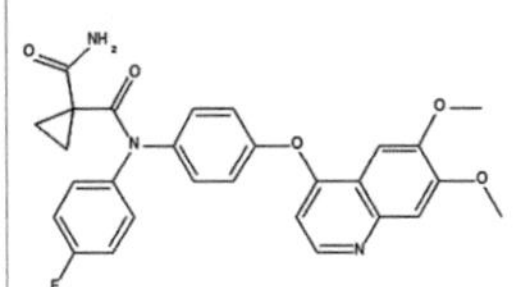	Inhibits VEGFR-2 and c-MET	Hepatocellular cancer, Thyroid cancer, Renal cancer	[19]
Axitinib	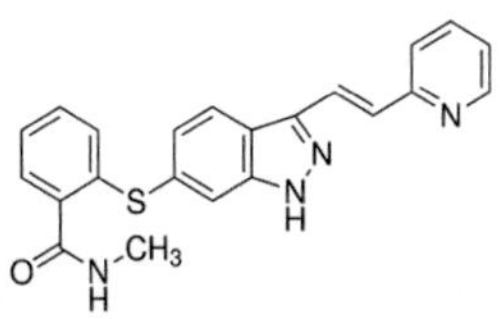	Inhibits the functions of VEGF-1, -2, and -3	Renal cell cancer	[20]
Regorafenib	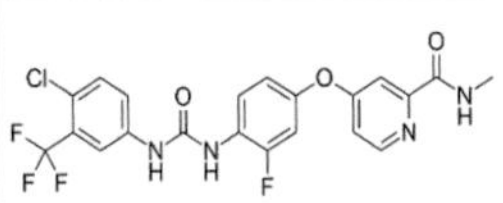	Inhibits VEGFR2-TIE2	Gastrointestinal stromal tumor, Colorectal cancer	[21]
Imatinib	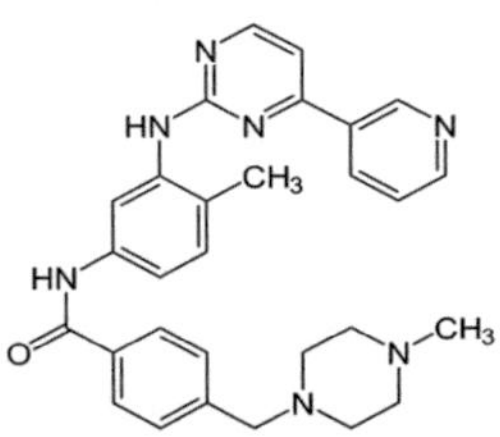	Tyrosine kinase inhibitor	Chronic myelogenous leukemia, Gastrointestinal stromal tumor	[22]
Everolimus	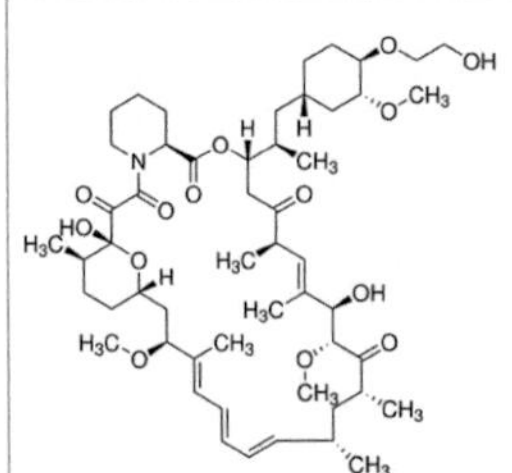	Immunosuppression and targets the mTOR pathway	Lung cancer, breast cancer	[23]

(continued)

Table 2.1 (continued)

Name	Structure	Function	Target cancer	References
Vandetanib	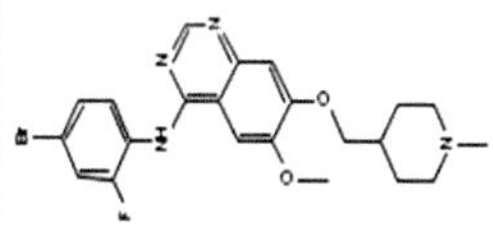	Inhibits both Epidermal Growth Factor Receptor (EGFR) and VEGFR Tyrosine Kinases	Medullary thyroid cancer	[24]
Ziv-aflibercept	A recombinant fusion protein containing the extracellular domains of human VEGF receptors 1 and 2	Inhibits VEGF	Colorectal cancer	[25]
Ramucirumab	Mmonoclonal antibody	Anti-VEGFR2	Non-small cell lung cancer, hepatocellular cancer	[26]
Denibulin (MN-029)	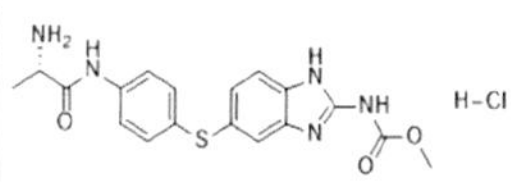	Vascular-disrupting agent (VDA), inhibitor of microtubule assembly	Solid tumors	[27]
Ombrabulin	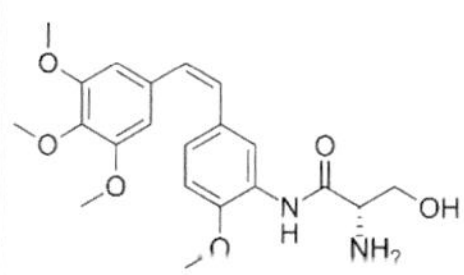	VDA	Head and neck squamous cancer	[9]

angiogenic cascades. Loges et al. showed different mechanisms of resistance in anti-angiogenesis based therapies with new proposals to improve the new generation drug candidates. Cancerous cells frequently play a role in producing drug resistance against the anti-angiogenic therapies across numerous ways [32].

Several of these processes are related to the tumor cells themselves, although cancerous cells may take over their microenvironment to accelerate the escape from VEGF-targeted therapies. Consequently, standardization of the microenvironment improves to utilize third age group anti-angiogenic drugs. Tumors can shift amid various modes of vascularization including vasculogenesis, sprouting angiogenesis, vascular mimicry and vessel cooption to confirm adequate nourishment to obtain drug resistance. It has been discovered that penetrability of the neovessels in glioblastoma (GBM) is irregular in nature compared to the common brain blood vessels causing in vasogenic edema. This irregular penetrability is mainly due to the atypical geomorphology of the neovascular 'front' throughout hypoxic state that assistances in the excretion of small peptides and different additional angiogenic factors. Moreover, cancerous cells can generate several pro-angiogenic factors/cytokines that assists in the amplification of pro-angiogenic genes that cause tumor progression and metastasis triggering multidrug resistance.

Apart from this, anti-angiogenic therapies itself might imply an escaping process by the upregulation of many growth factors accountable for liberating tumor vascularization [33]. For instance, VEGFR2 blockade can upregulate several pro-angiogenic components (FGF, VEGF, angiopoietin-1, PIDF, etc.) in tumor containing mice [34]. Likewise, throughout VEGF targeted therapy in various glioblastoma patients, plasma levels of FGF-2, SDF-1 have been observed at an elevated level during disease development [35]. Additional evade concepts involve the uncontrolled growth of angiogenesis autonomous and 'hypoxic resistant' malignant cells. Hypoxic cancer condition similarly improves to discharge numerous pro-angiogenic elements (G-CSF and SDF-1) by vessel trimming [36]. Except hypoxia, the release of 'cytokine storm' by nutritious tissue supports a 'pseudo-inflamed' condition that stimulates cancer advancement, eructation, and metastatic lodging [37]. Yet again, small amount of oxygen in the antagonistic hypoxic tumor situation support in tumor intrusiveness to a distant cancerous area. Crucially, GBM stem cells, can cause recurrence of metastasizing cells, and develop hypoxic lenient, which promotes self-restoration and provide the growth to the bulk tumor. Furthermore, hypoxic circumstances aid in the enrollment of endothelial progenitor cells (EPCs) to recruit bone marrow-derived circulating cells (RBCCs) near the cancer vasculature, endorsing tumor angiogenesis by the excretion of numerous pro-angiogenic cytokines [38].

The additional vital process for resistance and evade is the stimulation of the HGF/MET molecular pathway. HGF are primarily created by bone marrow-derived cells (BMDCs) [38]. Appearance of the HGF receptor c-Met in a mice xenograft cancer model was principally discovered in vascular endothelial cells but not in the malignant cells. This indicates a responsibility for HGF-fueled c-Met signaling cascades in endothelial cell lines to circumvent anti-VEGF therapies. Eckerich et al. showed that hypoxia may trigger c-Met manifestation in glioma cells to enhance SF/HGF-stimulated cells migration [39]. Obstruction of angiogenesis benefits in the growth of c-Met transcription and consequent infiltration of glioblastoma cancer cells. Hypoxic exposure promotes initiation of c-Met augmenting the HIF-1α amounts, causing cancer cell migration. VEGF as well play a crucial role to adversely regulate the functions of c-Met by the forming c-Met–VEGFR2 building complex on malignant cells through a hypoxia autonomous cascade [39].

One more important restriction of anti-angiogenic treatment linked with radiation is the reduction in the response to radiotherapy in lowered oxygen environment in a hypoxic condition. Fenton et al. exhibited that tumor hypoxia develops at a faraway place from the diffused blood vessels following 24 h of post radiation, indicating a reduction in oxygen utilization [40]. This may incidentally stimulate tumor endurance and cancer cell invasion. Therefore, tumor hypoxia may be responsible for the limitation of radiotherapy efficacy. Though, the fundamental physiological processes at the back of these limitations of radiotherapy in presence of hypoxic situations continue to be ambiguous that requires comprehensive systematic investigations.

Additional drawbacks consist of various side effects and drug toxicities of anti-angiogenic chemical drugs to different organs and tissues including long term toxicities, high-level blood pressure, threat of prenatal period and surgical treatment, hemorrhage or injury in the intestinal area. Lack of information regarding bioavailability, biodegradability, pharmacokinetics, pharmacodynamics etc. may need to be analyzed before a successful therapeutic strategy towards human.

References

1. K. Yoncheva, G. Momekov, Antiangiogenic anticancer strategy based on nanoparticulate systems, Expert Opin Drug Deliv 8(8) (2011) 1041–56.
2. Y. Cao, R. Langer, A review of Judah Folkman's remarkable achievements in biomedicine, Proc Natl Acad Sci U S A 105(36) (2008) 13203–5.
3. D. Hanahan, J. Folkman, Patterns and Emerging Mechanisms of the Angiogenic Switch during Tumorigenesis, Cell 86(3) (1996) 353–364.
4. D. Hanahan, R.A. Weinberg, Hallmarks of cancer: the next generation, Cell 144(5) (2011) 646–74.
5. N. Ferrara, Vascular Endothelial Growth Factor, Arteriosclerosis, Thrombosis, and Vascular Biology 29(6) (2009) 789–791.
6. N. Ferrara, Pathways mediating VEGF-independent tumor angiogenesis, Cytokine & Growth Factor Reviews 21(1) (2010) 21–26.
7. N. Ferrara, A.P. Adamis, Ten years of anti-vascular endothelial growth factor therapy, Nature Reviews Drug Discovery 15(6) (2016) 385–403.
8. G.W. Prager, M. Poettler, M. Unseld, C.C. Zielinski, Angiogenesis in cancer: Anti-VEGF escape mechanisms, Transl Lung Cancer Res 1(1) (2012) 14–25.
9. P. Hinnen, F.A. Eskens, Vascular disrupting agents in clinical development, Br J Cancer 96(8) (2007) 1159–65.
10. D. Ribatti, A. Vacca, New Insights in Anti-Angiogenesis in Multiple Myeloma, Int J Mol Sci 19(7) (2018).
11. J. Montoro, A. Yerlikaya, A. Ali, A. Raza, Improving Treatment for Myelodysplastic Syndromes Patients, Curr Treat Options Oncol 19(12) (2018) 66.
12. L.E. Minion, K.S. Tewari, The safety and efficacy of bevacizumab in the treatment of patients with recurrent or metastatic cervical cancer, Expert Rev Anticancer Ther 17(3) (2017) 191–198.
13. T.F. Cloughesy, A. Brenner, J.F. de Groot, N.A. Butowski, L. Zach, J.L. Campian, B.M. Ellingson, L.S. Freedman, Y.C. Cohen, N. Lowenton-Spier, T. Rachmilewitz Minei, S.F. Shmueli, G.S. Investigators, P.Y. Wen, A randomized controlled phase III study of VB-111 combined with bevacizumab vs. bevacizumab monotherapy in patients with recurrent glioblastoma (GLOBE), Neuro Oncol (2019).
14. Z. Hao, I. Sadek, Sunitinib: the antiangiogenic effects and beyond, Onco Targets Ther 9 (2016) 5495–505.
15. Z. Chen, H. Yang, Z. Li, Q. Xia, Y. Nie, Temsirolimus as a dual inhibitor of retinoblastoma and angiogenesis via targeting mTOR signalling, Biochem Biophys Res Commun 516(3) (2019) 726–732.
16. C.C. Cheng, W.T. Chao, C.C. Liao, J.H. Shih, Y.S. Lai, Y.H. Hsu, Y.H. Liu, The Roles Of Angiogenesis And Cancer Stem Cells In Sorafenib Drug Resistance In Hepatocellular Carcinoma, Onco Targets Ther 12 (2019) 8217–8227.

17. M. Matsuki, T. Hoshi, Y. Yamamoto, M. Ikemori-Kawada, Y. Minoshima, Y. Funahashi, J. Matsui, Lenvatinib inhibits angiogenesis and tumor fibroblast growth factor signaling pathways in human hepatocellular carcinoma models, Cancer Med 7(6) (2018) 2641–2653.
18. D.K. Chellappan, J. Chellian, Z.Y. Ng, Y.J. Sim, C.W. Theng, J. Ling, M. Wong, J.H. Foo, G.J. Yang, L.Y. Hang, S. Nathan, Y. Singh, G. Gupta, The role of pazopanib on tumour angiogenesis and in the management of cancers: A review, Biomed Pharmacother 96 (2017) 768–781.
19. A. Abdelaziz, U. Vaishampayan, Cabozantinib for the treatment of kidney cancer, Expert Rev Anticancer Ther 17(7) (2017) 577–584.
20. H. Kamli, L. Li, G.C. Gobe, Limitations to the Therapeutic Potential of Tyrosine Kinase Inhibitors and Alternative Therapies for Kidney Cancer, Ochsner J 19(2) (2019) 138–151.
21. J.B. Rey, V. Launay-Vacher, C. Tournigand, Regorafenib as a single-agent in the treatment of patients with gastrointestinal tumors: an overview for pharmacists, Target Oncol 10(2) (2015) 199–213.
22. C. Raimondi, A. Fantin, A. Lampropoulou, L. Denti, A. Chikh, C. Ruhrberg, Imatinib inhibits VEGF-independent angiogenesis by targeting neuropilin 1-dependent ABL1 activation in endothelial cells, J Exp Med 211(6) (2014) 1167–83.
23. L. Lee, T. Ito, R.T. Jensen, Everolimus in the treatment of neuroendocrine tumors: efficacy, side-effects, resistance, and factors affecting its place in the treatment sequence, Expert Opin Pharmacother 19(8) (2018) 909–928.
24. A. Morabito, M.C. Piccirillo, F. Falasconi, G. De Feo, A. Del Giudice, J. Bryce, M. Di Maio, E. De Maio, N. Normanno, F. Perrone, Vandetanib (ZD6474), a dual inhibitor of vascular endothelial growth factor receptor (VEGFR) and epidermal growth factor receptor (EGFR) tyrosine kinases: current status and future directions, Oncologist 14(4) (2009) 378–90.
25. J.I. Ivanova, K.R. Saverno, J. Sung, M.S. Duh, C. Zhao, S. Cai, F. Vekeman, A. Peevyhouse, R. Dhawan, C.S. Fuchs, Real-world treatment patterns and effectiveness among patients with metastatic colorectal cancer treated with ziv-aflibercept in community oncology practices in the USA, Med Oncol 34(12) (2017) 193.
26. M. Javle, E.C. Smyth, I. Chau, Ramucirumab: successfully targeting angiogenesis in gastric cancer, Clin Cancer Res 20(23) (2014) 5875–81.
27. A.D. Ricart, E.A. Ashton, M.M. Cooney, J. Sarantopoulos, J.M. Brell, M.A. Feldman, K.E. Ruby, K. Matsuda, M.S. Munsey, G. Medina, A. Zambito, A.W. Tolcher, S.C. Remick, A phase I study of MN-029 (denibulin), a novel vascular-disrupting agent, in patients with advanced solid tumors, Cancer Chemotherapy and Pharmacology 68(4) (2011) 959–970.
28. Y. Kubota, Tumor angiogenesis and anti-angiogenic therapy, Keio J Med 61(2) (2012) 47–56.
29. L. Moserle, G. Jimenez-Valerio, O. Casanovas, Antiangiogenic therapies: going beyond their limits, Cancer Discov 4(1) (2014) 31–41.
30. N. Dey, P. De, L.J. Brian, Evading anti-angiogenic therapy: resistance to anti-angiogenic therapy in solid tumors, Am J Transl Res 7(10) (2015) 1675–98.
31. N. Ferrara, R.S. Kerbel, Angiogenesis as a therapeutic target, Nature 438(7070) (2005) 967–74.
32. S. Loges, T. Schmidt, P. Carmeliet, Mechanisms of resistance to anti-angiogenic therapy and development of third-generation anti-angiogenic drug candidates, Genes Cancer 1(1) (2010) 12–25.
33. R.K. Jain, Normalization of tumor vasculature: an emerging concept in antiangiogenic therapy, Science 307(5706) (2005) 58–62.
34. M. Relf, S. LeJeune, P.A. Scott, S. Fox, K. Smith, R. Leek, A. Moghaddam, R. Whitehouse, R. Bicknell, A.L. Harris, Expression of the angiogenic factors vascular endothelial cell growth factor, acidic and basic fibroblast growth factor, tumor growth factor beta-1, platelet-derived endothelial cell growth factor, placenta growth factor, and pleiotrophin in human primary breast cancer and its relation to angiogenesis, Cancer Res 57(5) (1997) 963–9.

35. M. Paez-Ribes, E. Allen, J. Hudock, T. Takeda, H. Okuyama, F. Vinals, M. Inoue, G. Bergers, D. Hanahan, O. Casanovas, Antiangiogenic therapy elicits malignant progression of tumors to increased local invasion and distant metastasis, Cancer Cell 15(3) (2009) 220–31.
36. J.L. Yu, J.W. Rak, B.L. Coomber, D.J. Hicklin, R.S. Kerbel, Effect of p53 status on tumor response to antiangiogenic therapy, Science 295(5559) (2002) 1526–8.
37. O. Casanovas, D.J. Hicklin, G. Bergers, D. Hanahan, Drug resistance by evasion of antiangiogenic targeting of VEGF signaling in late-stage pancreatic islet tumors, Cancer Cell 8(4) (2005) 299–309.
38. K.V. Lu, G. Bergers, Mechanisms of evasive resistance to anti-VEGF therapy in glioblastoma, CNS Oncol 2(1) (2013) 49–65.
39. C. Eckerich, S. Zapf, R. Fillbrandt, S. Loges, M. Westphal, K. Lamszus, Hypoxia can induce c-Met expression in glioma cells and enhance SF/HGF-induced cell migration, Int J Cancer 121(2) (2007) 276–83.
40. B.M. Fenton, E.M. Lord, S.F. Paoni, Effects of radiation on tumor intravascular oxygenation, vascular configuration, development of hypoxia, and clonogenic survival, Radiat Res 155(2) (2001) 360–8.

3 Nanomedicine: An Alternative Approach Towards Anti-angiogenic Cancer Therapy

The molecular mediators of angiogenesis consist of different growth factors and cytokines (e.g., VEGF and FGF), matrix metalloproteinases (MMPs), and molecules involved in intracellular signaling pathways (Rho GTPases) [1]. There are specific types of receptors on the surface of cells (e.g., ECs) that act in response to angiogenic biomolecules; receptor tyrosine kinases (RTKs) are among the largest and most well-known receptor families [2]. VEGF receptors (VEGFR1–3), FGF receptors (FGFRs), PDGF receptors (PDGFRs), IGF receptors (IGFRs), and the Tie receptors (Tie1 and Tie2) are different classes of RTKs mediating angiogenesis through the activation of relevant signaling pathways after receiving the appropriate signals [3, 4]. For instance, the coupling of the IGF to its receptor (IGFRs) triggers two distinct signaling pathways in the cells, resulting from improved angiogenesis in the hypoxia condition [5]. Angiogenesis inhibitors can be divided into two distinct classes, including directly targeting the microvascular ECs, and those indirectly targeting the pro-angiogenic communication pathways between the cancer cells and ECs [6, 7]. A number of direct inhibitors (e.g., angiostatin) have been identified and used to inhibit angiogenesis in cancer treatment [8]. The main action of these inhibitors is to prevent the proliferation and migration of ECs stimulated by angiogenesis inducers (e.g., VEGF) [9]. An inhibitory effect on integrin receptors and subsequent signaling pathways is another mechanism proposed for the action of direct angiogenic inhibitors by which they prevent the proliferation of ECs. [1, 10, 11] The U.S. FDA has approved several angiogenesis inhibitors for the treatment of cancer [12, 13]. R. K. Jain reported that for both direct and indirect anti-angiogenic therapy, the balance between pro-angiogenic and anti-angiogenic factors could be restored by reducing vessel permeability and hypoxia and enhancing the homogeneity of blood flow perivascular cell coverage [14].

VEGF is a prevalent regulator of angiogenesis processes and is overexpressed in several human cancers [15]. Figure 3.1 shows that the VEGF-A, VEGF-B, VEGF-C, VEGF-D, and placental growth factor PIGF are the five VEGF family member ligands,

S. Mukherjee and V. S. Madamsetty, *Nanoparticles in Angiogenesis and Cancer*, Synthesis Lectures on Biomedical Engineering,
https://doi.org/10.1007/978-3-031-11284-3_3

which bind with three receptors (VEGFR-1, VEGFR-2, and VEGFR-3) [16, 17]. Endothelial cell migration and cell mitogenesis happen with the interaction of VEGF-A and VEGFR-2, which leads to cancer growth and metastasis [18]. The maintenance of the newly formed blood vessels occurs with the communication of VEGF-B and VEGFR-3. Further, VEGFR-3 and its ligands (VEGF-C and VEGF-D) play a crucial role in lymphangiogenesis and cancer cell migration to lymph nodes [19, 20]. Another VEGF family ligand, PIGF, is a cytokine that performs multiple functions in angiogenesis, like promoting tumor growth by stimulating stromal cells, bone marrow-derived endothelial progenitors, and myeloid cells [21]. In addition, Neuropilin 1 (NRP1) and neuropilin 2 (NRP-2) are also found as VEGF co-receptors that can induce signaling independently and through the activity of VEGFR2 [22, 23]. Other cell surface receptors, including integrins and growth factor receptors, can crosstalk with VEGF [24]. Figure 3.1 also describes the recently developed anti-angiogenic agents and their clinical trial status [16].

Angiogenesis inhibitors are designed to target existing tumor infiltrating blood vessels and inhibit the formation of newly formed blood vessels, thereby halting tumor metabolism and growth [25, 26]. Based on the inherent nature of the neo-angiogenesis in primary tumors, angiogenesis inhibitors might activate the destruction of immature angiogenic vessels or stop the angiogenic switch, thereby preventing vascular metastasis [10]. In conjunction with combination chemotherapy, angiogenic inhibitors may increase the anti-tumor potency to a significant level as these agents can halt the tumor progression but not eliminate them [27–29]. There are two kinds of angiogenesis inhibitors; the first group includes direct inhibitors, agents that block main angiogenesis proteins. The second group comprises indirect inhibitors that target the tumor cell or stromal cells and modulate angiogenesis regulators [30]. The direct angiogenesis inhibitors employ their anti-angiogenic effect by binding with angiogenesis inducers like VEGF, PDGF, and bFGF [31]. Bevacizumab, also known as Avastin, was the first FDA-approved anti-angiogenic therapy to treat several cancers [32]. Anti-angiogenic therapy-based treatment is the fourth cancer treatment modality, targeting tumor microenvironment microvascular endothelial cells and not the tumor cells [33, 34]. Fast-growing tumors such as melanoma and GBM need more vasculature and are more prone to anti-angiogenic therapy [34].

3.1 Limitations of Conventional Anti-angiogenic Therapy

Conventional anti-angiogenic therapy for cancer includes several limitations that can block a successful therapeutic strategy [35, 36]. Recently, the anti-angiogenic therapy by obstructing the VEGF pathway is the most important and widely accepted approach for the treatment of cancer; however, the clinical usage of this modality is still inadequate by several issues such as adverse effects, toxicity, acquired drug resistance, and non-availability of valid biomarkers [6]. Though Bevacizumab (Avastin), the first (VEGF) targeted anti-angiogenesis drug, was clinically approved for cancer therapy, the failure

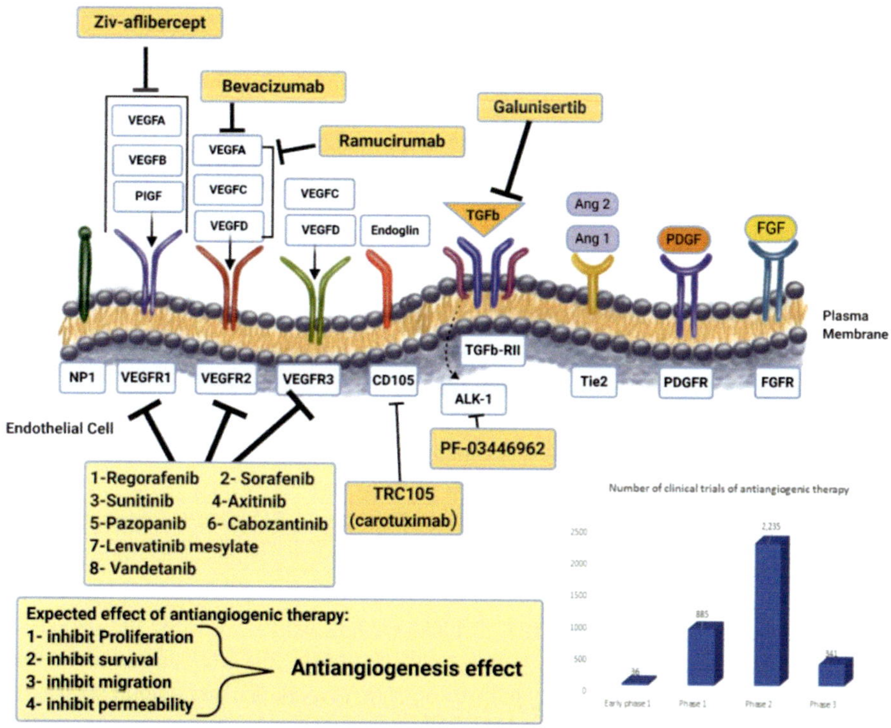

Fig. 3.1 Angiogenesis family member's ligands, receptors, and drug targets developed. VEGF-axis dependent and non-VEGF-mediated mechanisms of resistance to anti-angiogenic therapies. Non-VEGF axis receptors: TGF-β receptor, Tie2, PDGFR, and FGFR. Anti-angiogenic drug's mechanisms of action for anti-VEGF (monoclonal antibodies, RTK inhibitors) and novel targeted therapies are presented. Additionally, the chart represents the number of clinical trials of anti-angiogenic therapy at all different phases. Data obtained from https://www.clinicaltrials.gov/ (accessed on 4 February 2021 [16])

of this therapy is due to the development of inherent/the acquired resistance [37]. This major challenge leads to an increase in the understanding of VEGF independent angiogenesis [38]. Scientists have found that VEGF-A may be replaced by other VEGF family gene members like VEGF-C or VEGF-D, which bind to and activate the VEGFR [39]. Also, VEGF-A may be replaced by a different angiogenic cascade [40]. Scientists, nicely demonstrated the various resistance mechanisms to anti-angiogenic therapy and proposed improving third-generation anti-angiogenic drug candidates [41]. Malignant cells often generate resistance against the anti-angiogenic treatment in several ways [42]. Some of these mechanisms are attributable to the tumor cells themselves, but malignant cells also seem to highjack their microenvironment to stimulate escape from VEGF-targeted therapy [43]. Hence normalization of the microenvironment helps to develop third-generation of anti-angiogenic drugs [41]. Tumors can switch between different modes of vascularization such as sprouting angiogenesis, vasculogenesis, vessel cooption, and vascular mimicry to ensure sufficient nutrition to acquire drug resistance [44]. It has been found that the permeability of neovessels in glioblastoma (GBM) is abnormal compared to the normal brain blood vessels resulting in vasogenic edema [45]. This abnormal permeability of brain blood neovessels is due to the abnormal morphology of the neovascular front during hypoxic conditions, which helps in the secretion of small peptides and other angiogenic fact [46]. Besides this, malignant cells can produce multiple pro-angiogenic factors/cytokines and amplify the genes associated with angiogenesis during the tumor progression and metastasis [47], resulting in the acquisition of drug resistance [48]. Besides this, anti-angiogenic therapy may suggest an escape mechanism by upregulating various growth factors responsible for rescuing tumor vascularization [49]. For example, in tumor bearing mice, VEGFR2 blockade upregulates various pro-angiogenic factors (VEGF, FGF, PIDF, angiopoietin-1, and others) [44].

Similarly, during VEGF, targeted therapy in glioblastoma, which is one of the most aggressive angiogenic tumors, plasma levels of FGF-2 and SDF-1 are found to increase under disease progression [50]. Apart from this, several preclinical and clinical studies demonstrated the enhanced invasiveness and metastasis of glioblastoma upon blockade of VEGF using inhibitors or drugs [51]. Other escaping theories include the overgrowth of hypoxic resistant and angiogenesis independent cancer cells [52]. It has to be noted that hypoxic tumor conditions also help to secrete several pro-angiogenic factors (G-CSF and SDF-1) by vessel pruning [48]. Other than hypoxia, the release of cytokine storm by healthy tissue also promotes an n pseudo-inflammed state that induces tumor progression, extravasations, and metastatic lodging [53]. Importantly, glioblastoma (GBM) cancer stem cells (CSC), which help in the recurrence of metastasizing cancer cells, can become hypoxic tolerant and resides in hypoxic tumor regions, which can sustain self-renewal and maintain the bulk tumor [54]. Additionally, hypoxic conditions helped in the recruitment of endothelial progenitor cells (EPCs) and recruited bone marrow-derived circulating cells (RBCCs) towards the tumor vasculature, which can indirectly promote tumor angiogenesis by the secretion of several angiogenic cytokines [55].

3.2 Alternative Anti-angiogenic Therapy: Nanomedicine Based Approach

The angiogenesis research revolutionized the development of anti-angiogenic drugs that inhibit cancer cell growth, useful for cancer treatment [56]. The anti-angiogenic approach inhibits either the binding process between pro-angiogenic growth factors and their corresponding receptors or the activity of proteolytic enzymes of the extracellular matrix [41]. The toxicity of most anti-angiogenic drug molecules (small molecules) or instability of anti-angiogenic proteins requires developing a new formulation in an appropriate delivery system [41]. In this context, nanomedicine plays a pivotal role. There has been considerable progress in developing anti-cancer therapeutics that explicitly act on detectable abnormalities in tumors, minimizing normal cell damage [57]. The type of cancer impacts the prognosis for patients; however, the grading and the occurrence of specific molecular markers can also lead to establishing the prognosis and controlling individual treatments [57]. Here we will focus on the recent advances in nanomaterials (i.e., nanoparticles produced by biomaterials) used to develop innovative approaches for anti-angiogenic oncology. Nanomaterials may be utilized to treat cancer either by passive or active pathways. In the passive pathway, nanomaterials circulate in the bloodstream and are accumulated into the tumor through enhanced permeability and retention effect (EPR effect); in the active pathway, nanoparticles reach the tumor through a receptor-targeted way [58]. Several materials, such as lipids, phospholipids, polymers, etc., will be used to produce these nanoparticles, with specific size and surface charge. Further, these can even be attached with specific targeting ligands on the surface for site-specific targeting. Notably, numerous nanomaterials, including liposomes, lipid-polymeric NPs, polymer NPs, and protein NPs, are utilized to target anti-angiogenic agents to tumors to control tumor angiogenesis (Fig. 3.2) [26].

Additionally, nanomedicines are being investigated for their use in anti-cancer therapies to improve drug delivery, increase treatment efficacy, reduce side effects, and overcome drug resistance [59]. The number of studies published under the research topics of nanomedicine and nanotechnology has increased potentially over the past decade. Currently, the knowledge base of nanoparticles is still intensifying, highlighting safety and efficacy. The unique properties of nanomaterials, such as small size, high surface-to-volume ratio, and high surface energy compared to bulk materials, can be used to deliver the active components to target sites [60]. Nanotechnology is a platform where researchers from different backgrounds of chemists, pharmacologists, physicists, biologists, medical doctors, materials science scientists, and engineers can work together for interdisciplinary and integrative research [61]. The successful translational of nanotechnology-based therapeutic and diagnostic tools from laboratory to clinic and vice-versa needs extensive multidisciplinary research [62]. Several scientists, including our group, recently showed the nanotechnology-based imaging and therapeutic approach using angiogenic markers for treating cancer. In general, fabricated nanomaterials can overcome the limitations found in conventional treatment approaches [63]. Sometimes, active anti-angiogenic agents are directly conjugated to nanoparticles containing targeting ligands which can improve the therapeutic efficacy of anti-angiogenic agents in the tumor microenvironment [64]. Since conventional anti-angiogenic treatment includes various limitations, we believe that the nanomedicine method could be used as an alternative treatment strategy for cancers using the anti-angiogenic properties of nanomaterials or nanoparticle-based drug delivery systems. Recent research reports reveal both pro-and anti-angiogenic properties of nanomaterials [11]. Several groups, including ours, have shown the pro-angiogenic properties

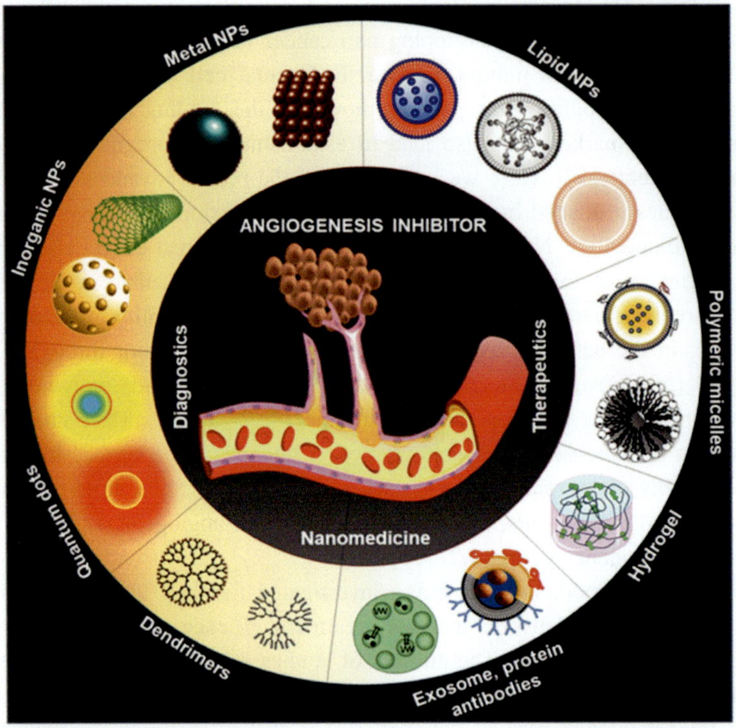

Fig. 3.2 Schematic representation of multiple nanomedicine approaches that may be used for the diagnosis and treatment of angiogenesis [26]

of different nanomaterials, including liposomes, copper nanoparticles (CuNPs), europium hydroxide nanorods (EHNs), zinc oxide nanoflowers, carbon nanotubes (CNW), graphene oxides (GO), cerium oxide nanoparticles [35].

Figure 3.3 demonstrates that angiogenesis-based nanomaterials used for diagnosis purposes by using positron emission tomography (PET) and single-photon emission computed tomography (SPECT) are considerably noninvasive imaging modalities in clinical settings for oncology [16].

On the other hand, nanoparticle-based anti-angiogenic drug delivery systems were developed to suppress tumor angiogenesis [36, 65–67]. Furthermore, the unique properties of nanoparticulate systems (nanoparticles, liposomes, polymeric micelles, etc.) can be used to target endothelial cells and the efficient delivery of anti-angiogenic agents [68]. Various nanoparticles show the targeting ability. Nanoparticles conjugated with the corresponding targeting agent can target VEGF and its receptors, fibroblast growth factor, and its receptors, EGFRs, MMPs, tubulin function, etc. [69]. These targeted drug delivery systems allow them to deliver both anti-angiogenic molecules and anti-cancer drugs,

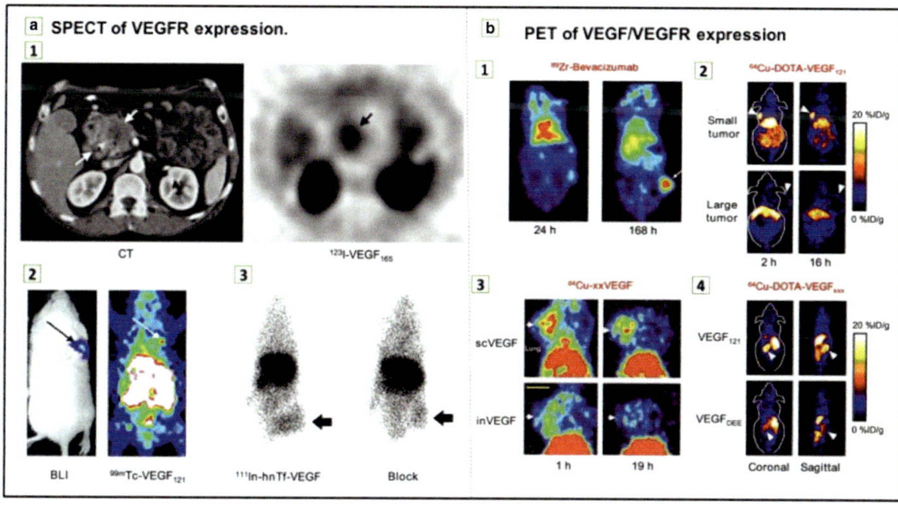

Fig. 3.3 **a** Single-photon emission computed tomography (SPECT) of vascular endothelial growth factor receptor (VEGFR) expression. (1) Transverse CT image of pancreatic adenocarcinoma patient (left) and transverse SPECT image of the same patient at 1.5 h after injection of 123I-VEGF165 (right). (2) Bioluminescence imaging (BLI; after injection of d-luciferin) and SPECT images (after injection of 99mTc-VEGF121) of a tumor-bearing mouse. Tumor cells were transfected with firefly luciferase. (3) Posterior whole-body images of the tumor-bearing mouse at 48 h after injection of 111In-hnTf-VEGF and after coinjection of 100-fold excess of unlabeled apotransferrin (block). Arrows in all images indicate tumors. **b** Positron emission tomography (PET) of VEGF/VEGFR expression. (1) Coronal small-animal PET images of a tumor-bearing mouse at 24 and 168 h after injection of 89Zr-bevacizumab. (2) Coronal small-animal PET images of U87MG tumor-bearing mice at 2 and 16 h after injection of 64Cu-DOTA-VEGF121. The small tumor expressed a high level of VEGFR-2, and the large tumor expressed a low level of VEGFR-2. (3) Coronal small-animal PET images of 4T1 tumor-bearing mice at 1 and 19 h after injection of either 64Cu-scVEGF (single-chain VEGF that binds to VEGFR) or the equivalent amount of 64Cu-in VEGF (inactive VEGF that does not bind to VEGFR). (4) Coronal and sagittal slices containing kidneys (arrowheads) at 4 h after injection of 64Cu-DOTA-VEGF121 (binds to both VEGFR-1 and VEGFR-2) or 64Cu-DOTA-VEGFDEE (VEGFR-2-specific). Arrows in 1–3 indicate tumors [16]

facilitate drug penetration into extravascular tumor tissue and consequently increase the therapeutic efficacy and reduce the systemic toxicity. In summary, nanomedicine-based anti-angiogenic therapy may be an excellent alternative approach to the clinic by replacing current conventional anti-angiogenic therapies.

References

1. A.A. Ucuzian, A.A. Gassman, A.T. East, H.P. Greisler, Molecular mediators of angiogenesis, Journal of Burn Care & Research 31(1) (2010) 158–175.
2. A. Papa, E. Zaccarelli, D. Caruso, P. Vici, P. Benedetti Panici, F. Tomao, Targeting angiogenesis in endometrial cancer-new agents for tailored treatments, Expert Opinion on Investigational Drugs 25(1) (2016) 31–49.
3. M. Shibuya, Vascular Endothelial Growth Factor (VEGF) and Its Receptor (VEGFR) Signaling in Angiogenesis: A Crucial Target for Anti- and Pro-Angiogenic Therapies, Genes Cancer 2(12) (2011) 1097–1105.
4. F.W. Liu, J. Cripe, K.S. Tewari, Anti-angiogenesis therapy in gynecologic malignancies, Oncology 29(5) (2015) 350–350.
5. S. Kargozar, F. Baino, S. Hamzehlou, M.R. Hamblin, M. Mozafari, Nanotechnology for angiogenesis: opportunities and challenges, Chemical Society Reviews 49(14) (2020) 5008–5057.
6. R.N. Gacche, R.J. Meshram, Angiogenic factors as potential drug target: efficacy and limitations of anti-angiogenic therapy, Biochimica et Biophysica Acta (BBA)-Reviews on Cancer 1846(1) (2014) 161–179.
7. R. Ronca, M. Benkheil, S. Mitola, S. Struyf, S. Liekens, Tumor angiogenesis revisited: regulators and clinical implications, Medicinal research reviews 37(6) (2017) 1231–1274.
8. J. Folkman, Angiogenesis inhibitors: a new class of drugs, Cancer biology & therapy 2(sup1) (2003) 126–132.
9. S.Y. Yoo, S.M. Kwon, Angiogenesis and its therapeutic opportunities, Mediators of inflammation 2013 (2013).
10. B. Döme, M.J. Hendrix, S. Paku, J. Tóvári, J. Tímár, Alternative vascularization mechanisms in cancer: Pathology and therapeutic implications, The American journal of pathology 170(1) (2007) 1–15.
11. R. Giordo, Z. Wehbe, P. Paliogiannis, A.H. Eid, A.A. Mangoni, G. Pintus, Nano-targeting Vascular Remodelling in Cancer: Recent Developments and Future Directions, Seminars in Cancer Biology, Elsevier, 2022.
12. S. Qin, A. Li, M. Yi, S. Yu, M. Zhang, K. Wu, Recent advances on anti-angiogenesis receptor tyrosine kinase inhibitors in cancer therapy, Journal of hematology & oncology 12(1) (2019) 1–11.
13. N.A. Shukla, M.N. Yan, N. Hanna, The story of angiogenesis inhibitors in non–small-cell lung cancer: the past, present, and future, Clinical Lung Cancer 21(4) (2020) 308–313.
14. S. Goel, D.G. Duda, L. Xu, L.L. Munn, Y. Boucher, D. Fukumura, R.K. Jain, Normalization of the vasculature for treatment of cancer and other diseases, Physiological reviews 91(3) (2011) 1071–1121.
15. Y. Yang, Y. Cao, The impact of VEGF on cancer metastasis and systemic disease, Seminars in Cancer Biology, Elsevier, 2022.
16. H.O. Alsaab, A.S. Al-Hibs, R. Alzhrani, K.K. Alrabighi, A. Alqathama, A. Alwithenani, A.H. Almalki, Y.S. Althobaiti, Nanomaterials for antiangiogenic therapies for cancer: a promising tool for personalized medicine, International Journal of Molecular Sciences 22(4) (2021) 1631.
17. S.A. Park, M.S. Jeong, K.-T. Ha, S.B. Jang, Structure and function of vascular endothelial growth factor and its receptor system, BMB reports 51(2) (2018) 73.
18. C. Ceci, M.G. Atzori, P.M. Lacal, G. Graziani, Role of VEGFs/VEGFR-1 signaling and its inhibition in modulating tumor invasion: Experimental evidence in different metastatic cancer models, International journal of molecular sciences 21(4) (2020) 1388.

19. S. Jha, Mechanism of VEGF-C Activation and Effect on Lymphatic Vessel Growth and Regeneration, (2020).
20. D. Ribatti, B. Nico, E. Crivellato, A. Roccaro, A. Vacca, The history of the angiogenic switch concept, Leukemia 21(1) (2007) 44–52.
21. N. Ferrara, Pathways mediating VEGF-independent tumor angiogenesis, Cytokine & growth factor reviews 21(1) (2010) 21–26.
22. C. Hu, X. Jiang, Role of NRP-1 in VEGF-VEGFR2-independent tumorigenesis, Targeted oncology 11(4) (2016) 501–505.
23. H.L. Goel, A.M. Mercurio, VEGF targets the tumour cell, Nature Reviews Cancer 13(12) (2013) 871–882.
24. P.R. Somanath, A. Ciocea, T.V. Byzova, Integrin and growth factor receptor alliance in angiogenesis, Cell biochemistry and biophysics 53(2) (2009) 53–64.
25. N.M. Pandya, N.S. Dhalla, D.D. Santani, Angiogenesis—a new target for future therapy, Vascular pharmacology 44(5) (2006) 265–274.
26. A. Mukherjee, V.S. Madamsetty, M.K. Paul, S. Mukherjee, Recent advancements of nanomedicine towards antiangiogenic therapy in cancer, International Journal of Molecular Sciences 21(2) (2020) 455.
27. R.S. Kerbel, Tumor angiogenesis: past, present and the near future, Carcinogenesis 21(3) (2000) 505–515.
28. V. Lai, S.Y. Neshat, A. Rakoski, J. Pitingolo, J.C. Doloff, Drug delivery strategies in maximizing anti-angiogenesis and anti-tumor immunity, Advanced Drug Delivery Reviews 179 (2021) 113920.
29. M.J. Ansari, D. Bokov, A. Markov, A.T. Jalil, M.N. Shalaby, W. Suksatan, S. Chupradit, H.S. Al-Ghamdi, N. Shomali, A. Zamani, Cancer combination therapies by angiogenesis inhibitors; a comprehensive review, Cell Communication and Signaling 20(1) (2022) 1–23.
30. M. Taleb, N. Mohammadkhani, F. Bahreini, M. Ovais, G. Nie, Modulation of Tumor Vasculature Network: Key Strategies, Small Structures (2022) 2100164.
31. F.H. Al-Ostoot, S. Salah, H.A. Khamees, S.A. Khanum, Tumor angiogenesis: Current challenges and therapeutic opportunities, Cancer Treatment and Research Communications 28 (2021) 100422.
32. E. Ioannidou, M. Moschetta, S. Shah, J.S. Parker, M.A. Ozturk, G. Pappas-Gogos, M. Sheriff, E. Rassy, S. Boussios, Angiogenesis and anti-angiogenic treatment in prostate cancer: mechanisms of action and molecular targets, International Journal of Molecular Sciences 22(18) (2021) 9926.
33. Y. Cao, Future options of anti-angiogenic cancer therapy, Chinese journal of cancer 35(1) (2016) 1–9.
34. A. Abdollahi, J. Folkman, Evading tumor evasion: current concepts and perspectives of anti-angiogenic cancer therapy, Drug Resistance Updates 13(1-2) (2010) 16–28.
35. S. Mukherjee, C.R. Patra, Therapeutic application of anti-angiogenic nanomaterials in cancers, Nanoscale 8(25) (2016) 12444–12470.
36. P. Liang, B. Ballou, X. Lv, W. Si, M.P. Bruchez, W. Huang, X. Dong, Monotherapy and combination therapy using anti-angiogenic nanoagents to fight cancer, Advanced Materials 33(15) (2021) 2005155.
37. Y. Itatani, K. Kawada, T. Yamamoto, Y. Sakai, Resistance to Anti-Angiogenic Therapy in Cancer-Alterations to Anti-VEGF Pathway, Int J Mol Sci 19(4) (2018).
38. F. Lopes-Coelho, F. Martins, S.A. Pereira, J. Serpa, Anti-angiogenic therapy: Current challenges and future perspectives, International Journal of Molecular Sciences 22(7) (2021) 3765.

39. J. Künnapuu, H. Bokharaie, M. Jeltsch, Proteolytic cleavages in the VEGF family: Generating diversity among angiogenic VEGFs, essential for the activation of lymphangiogenic VEGFs, Biology 10(2) (2021) 167.
40. S.J. Harper, D.O. Bates, VEGF-A splicing: the key to anti-angiogenic therapeutics?, Nature Reviews Cancer 8(11) (2008) 880–887.
41. A.M. Abdalla, L. Xiao, M.W. Ullah, M. Yu, C. Ouyang, G. Yang, Current challenges of cancer anti-angiogenic therapy and the promise of nanotherapeutics, Theranostics 8(2) (2018) 533.
42. G. Lupo, N. Caporarello, M. Olivieri, M. Cristaldi, C. Motta, V. Bramanti, R. Avola, M. Salmeri, F. Nicoletti, C.D. Anfuso, Anti-angiogenic therapy in cancer: downsides and new pivots for precision medicine, Frontiers in pharmacology 7 (2017) 519.
43. S. Loges, T. Schmidt, P. Carmeliet, Mechanisms of resistance to anti-angiogenic therapy and development of third-generation anti-angiogenic drug candidates, Genes Cancer 1(1) (2010) 12–25.
44. R. Lugano, M. Ramachandran, A. Dimberg, Tumor angiogenesis: causes, consequences, challenges and opportunities, Cellular and Molecular Life Sciences 77(9) (2020) 1745–1770.
45. M. Caffo, S.M. Cardali, E. Fazzari, V. Barresi, G. Caruso, Nanoparticles drug-delivery systems and antiangiogenic approaches in the treatment of gliomas, Glioma 1(6) (2018) 183.
46. J. Jászai, M.H. Schmidt, Trends and challenges in tumor anti-angiogenic therapies, Cells 8(9) (2019) 1102.
47. B. Bassani, D. Baci, M. Gallazzi, A. Poggi, A. Bruno, L. Mortara, Natural killer cells as key players of tumor progression and angiogenesis: old and novel tools to divert their pro-tumor activities into potent anti-tumor effects, Cancers 11(4) (2019) 461.
48. Y. Haibe, M. Kreidieh, H. El Hajj, I. Khalifeh, D. Mukherji, S. Temraz, A. Shamseddine, Resistance mechanisms to anti-angiogenic therapies in cancer, Frontiers in oncology 10 (2020) 221.
49. R.N. Gacche, Y.G. Assaraf, Redundant angiogenic signaling and tumor drug resistance, Drug Resistance Updates 36 (2018) 47–76.
50. S. Takano, Glioblastoma angiogenesis: VEGF resistance solutions and new strategies based on molecular mechanisms of tumor vessel formation, Brain tumor pathology 29(2) (2012) 73–86.
51. J.M. Ebos, R.S. Kerbel, Antiangiogenic therapy: impact on invasion, disease progression, and metastasis, Nature reviews Clinical oncology 8(4) (2011) 210–221.
52. P. Gaur, D. Bose, S. Samuel, L.M. Ellis, Targeting tumor angiogenesis, Seminars in oncology, Elsevier, 2009, pp. S12–S19.
53. N. Bassani, New targets in tumor angiogenesis to block tumor re-growth and therapeutic resistance, (2017).
54. C. Peitzsch, A. Tyutyunnykova, K. Pantel, A. Dubrovska, Cancer stem cells: The root of tumor recurrence and metastases, Seminars in cancer biology, Elsevier, 2017, pp. 10–24.
55. M. Furuya, Y. Yonemitsu, I. Aoki, III. Angiogenesis: complexity of tumor vasculature and microenvironment, Current pharmaceutical design 15(16) (2009) 1854–1867.
56. Z.R. Huinen, E.J. Huijbers, J.R. van Beijnum, P. Nowak-Sliwinska, A.W. Griffioen, Anti-angiogenic agents—overcoming tumour endothelial cell anergy and improving immunotherapy outcomes, Nature Reviews Clinical Oncology 18(8) (2021) 527–540.
57. R.I. Teleanu, C. Chircov, A.M. Grumezescu, D.M. Teleanu, Tumor angiogenesis and anti-angiogenic strategies for cancer treatment, Journal of clinical medicine 9(1) (2019) 84.

58. J.K. Tee, L.X. Yip, E.S. Tan, S. Santitewagun, A. Prasath, P.C. Ke, H.K. Ho, D.T. Leong, Nanoparticles' interactions with vasculature in diseases, Chemical Society Reviews 48(21) (2019) 5381–5407.
59. M.Z. Ahmad, M. Rizwanullah, J. Ahmad, M.Y. Alasmary, M.H. Akhter, B.A. Abdel-Wahab, M.H. Warsi, A. Haque, Progress in nanomedicine-based drug delivery in designing of chitosan nanoparticles for cancer therapy, International Journal of Polymeric Materials and Polymeric Biomaterials 71(8) (2022) 602–623.
60. A.B. Asha, R. Narain, Nanomaterials properties, Polymer Science and Nanotechnology, Elsevier2020, pp. 343–359.
61. H. Brune, H. Ernst, A. Grunwald, W. Grünwald, H. Hofmann, H. Krug, P. Janich, M. Mayor, W. Rathgeber, G. Schmid, Nanotechnology: Assessment and perspectives, Springer Science & Business Media2006.
62. S. Sahoo, S. Parveen, J. Panda, The present and future of nanotechnology in human health care, Nanomedicine: Nanotechnology, biology and medicine 3(1) (2007) 20–31.
63. Z. Cheng, M. Li, R. Dey, Y. Chen, Nanomaterials for cancer therapy: current progress and perspectives, Journal of Hematology & Oncology 14(1) (2021) 1–27.
64. E. Segal, R. Satchi-Fainaro, Design and development of polymer conjugates as anti-angiogenic agents, Advanced drug delivery reviews 61(13) (2009) 1159–1176.
65. J. Li, D.J. Burgess, Nanomedicine-based drug delivery towards tumor biological and immunological microenvironment, Acta Pharmaceutica Sinica B 10(11) (2020) 2110–2124.
66. Q. Liang, L. Zhou, Y. Li, J. Liu, Y. Liu, Nano drug delivery system reconstruct tumour vasculature for the tumour vascular normalisation, Journal of Drug Targeting 30(2) (2022) 119–130.
67. A.S. Oguntade, F. Al-Amodi, A. Alrumayh, M. Alobaida, M. Bwalya, Anti-angiogenesis in cancer therapeutics: The magic bullet, Journal of the Egyptian National Cancer Institute 33(1) (2021) 1–11.
68. J. Cao, D. Huang, N.A. Peppas, Advanced engineered nanoparticulate platforms to address key biological barriers for delivering chemotherapeutic agents to target sites, Advanced Drug Delivery Reviews 167 (2020) 170–188.
69. M.U. Rehman, A. Khan, Z. Imtiyaz, S. Ali, H.A. Makeen, S. Rashid, A. Arafah, Current Nanotherapeutic Approaches Ameliorating Inflammation in Cancer Progression, Seminars in cancer biology, Elsevier, 2022.

4 Organic Nanoparticles in Anti-angiogenic Cancer Therapy

Cancer is still playing significant global health concern despite recent remarkable advances in the research field [1–3]. Alterations in the vasculature network and upregulated angiogenesis are other main reasons for cancer progression and metastasis [4]. Angiogenesis is the development of new blood vessels from existing blood vessels through the proliferation and migration of discerned endothelial cells [5]. Angiogenesis is a key process engaged in embryogenesis and maintaining normal homeostasis, including repairing and regenerating injured tissues [6]. Angiogenesis may be deregulated in many pathological environments [7]. Although angiogenesis remains inactive during adulthood, it becomes physiologically active in normal situations such as the cycling ovary and the placenta during pregnancy [8]. Furthermore, angiogenesis consistently occurs via the activation of ECs in response to special stimuli (e.g., hypoxia) arising during the wound healing procedure to accelerate tissue reconstruction [9]. However, another story about unwanted angiogenesis occurs in many diseases and disorders, including cancer, i.e., an discrepancy between angiogenic stimulators and inhibitors indicates to triggering an angiogenic on-and-off switch [10]. For instance, the angiogenesis process is switched on in the case of malignancies and some inflammatory disorders [11]. On the contrary, insufficient angiogenesis is observed in other pathological conditions such as ischaemic heart tissue, in which healing and regeneration are impaired due to dysfunction of ECs, and vessel malformation or regression [12].

Recent clinical analysis evidenced that molecular angiogenesis targets play an important role in the early detection of cancer [13, 14]. Additionally, some most recent studies also showed that angiogenesis inhibitors, combined with other treatment regimens, improve the efficacy and prolong patient survival [14]. The current interest in nanoparticle delivery platforms is anticipated to enhance the cellular delivery of newer anti-angiogenic cancer therapeutics, as angiogenesis is crucial in tumor biology [15]. The angiogenic switch by vascular endothelial growth factor (VEGF) and vascular endothelial

S. Mukherjee and V. S. Madamsetty, *Nanoparticles in Angiogenesis and Cancer*, Synthesis Lectures on Biomedical Engineering,
https://doi.org/10.1007/978-3-031-11284-3_4

growth factor receptor (VEGFR) expression upregulation drives tumor blood vessel viability and cell survival, mainly triggered by tumor hypoxia [16]. The first preclinical model of anti-angiogenic therapy was developed by Kim et al. to demonstrate that the treatment of VEGF monoclonal antibody inhibited tumor growth [17]. These findings guided the development of VEGF/VEGFR signaling cascade target drugs [18]. Bevacizumab (sold under the brand name Avastin), VEGF monoclonal antibody, was the first anti-angiogenic drug approved to treat cancers [19]. Further, recently it is also approved for treatment in the combination of chemotherapy in some metastatic and advanced malignancies [20]. However, despite the significant success of using these anti-angiogenic drugs in the clinic, they still limit their usage due to treatment resistance and severe toxicity [21].

Nanoparticles are defined as substances where at least one of their dimensions is between 1 and 200 nm [22]. The application covers a wide interdisciplinary research development area that expands explosively worldwide [23]. Nanomaterials can transform into functionalized substitutes that can be further reaccessed [24]. Among several nanomaterials, organic-based nanomaterials are the most explored, which developed by using organic materials such as lipids, polymers, etc. [9]. We will discuss the different anti-angiogenic drugs based on organic nanomaterials and their applications. We will also highlight the recent advances in the field and their clinical applications.

4.1 Liposomes

Liposomes are spherical vesicles comprising phospholipids that may be either uni-lamellar or multi-lamellar to encapsulate hydrophobic, hydrophilic drug molecules and genes [25, 26]. This unique architecture makes it possible to entrap both hydrophilic and hydrophobic agents to effectively protect the trapped drug from environmental degradation [27]. Liposomes have characteristics such as biodegradability and biocompatibility, nontoxic, and less immunogenicity [28]. Liposomes are the first nano based drug that was approved in 1965, which is Doxil, a PEGylated dox liposome [29]. A typical liposome structure comprises a hydrophilic core and a hydrophobic phospholipid bilayer [30]. Liposomes offer an excellent platform for drug delivery such as doxorubicin, paclitaxel, and nucleic acid by demonstrating higher anti-tumor efficacy and enhanced bioavailability [31]. Doxil® and Myocet® are approved liposome-based nanoparticles of daunorubicin used to treat MBC [29, 32]. However, due to inadequacies like reduced encapsulation efficacy, rapid removal by MP, cell adsorption, and little shelf life, the application of liposome-based NPs is limited [33]. Rauca et al. 2021, assessed the effects of this novel combined tumor-targeted treatment on s.c. B16.F10 murine melanoma growth and the production of critical markers involved in tumor development and progression [34]. The results showed that the combined liposomal therapy almost totally inhibited (>90%) the growth of melanoma tumors due to the enhancement of anti-angiogenic effects of long-circulating liposomes-5, 6-dimethylxanthenone-4-acetic acid (LCL-DMXAA) by long-circulating liposomes-simvastatin (LCL-SIM) and suggest that novel therapy holds the potential to remodel the TME, by suppressing its most important malignant biological capabilities [34].

Another group of researchers synthesized surface-functionalize PEGylated liposomal doxorubicin (PLD) using an anti-p32 CGKRK peptide to evaluate its anti-angiogenic and anti-tumor activities [35]. CGKRK was conjugated to DSPE-mPEG2000-maleimide and post-inserted into PLD and concluded that PLD-CGKRK- could efficiently target endothelial and tumor parenchymal cells, which enhances the therapeutic efficacy of PLD and merits further investigation [35]. Renault-Mahieux et al. 2021, developed fisetin and cisplatin co-encapsulated liposomes. Increasing ratios of cholesterol/DOPC were used in preparing these liposomes and characterized in terms of size, PDI, and stability [36]. The integration of fisetin was explored using DSC. The anti-angiogenic and cytotoxic activities of the selected formulation were assayed in vitro, and successfully developed an optimized liposomal formulation incorporating both drugs [36]. The formulation demonstrated an additive influence of cisplatin and fisetin on GBM cells. The formulation was able to retain the activity of fisetin, was effective against GBM cells, and is promising for further in vivo experimentations [36].

The novel dual-targeted approach via cationic TSL (CTSL) was thoroughly explored by Petrini et al. 2021, using either a novel synthetic phospholipid 1,2-dipalmitoyl-sn-glycero-3-phosphodiglycerol (DPPG2) or a conventional polyethylene glycol (PEG) surface modification [37]. The authors concluded that synthesized cationic particles showed lower hemocompatibility than their anionic counterparts, which was partially mitigated when PEG was grafted on the liposome surface [37]. DPPG2-based anionic TSL showed optimal local drug delivery compared to all other formulations tested, demonstrating the potential advantages of using DPPG2 lipid in designing liposomes for tumor-targeted applications [37]. As shown in schematic Fig. 4.1, GA & Gal dual-ligand-modified liposomes co-loaded with curcumin and combretastatin A-4 phosphate (CUCA/GA&Gal-Lip) were prepared and characterized by Wang et al. 2021 [38]. The result demonstrated that dual-ligand-modified liposomes showed more significant inhibition of tumor angiogenesis and metastasis in comparison with other combined groups. The CUCA/GA&Gal-Lip holds excellent potential in hepatoma-targeting delivery of anti-tumor drugs. It can achieve anti-angiogenic and anti-metastatic effects by simultaneously blocking the VEGF/VEGFR2 signal pathway, exhibiting superior anti-hepatoma efficacy [38].

Another group of authors engineered a PEGylated liposomal system encapsulating lipocalin 2 (Lcn2) small interfering RNA (Lcn2 siRNA) for selective targeting of MBC cell line MCF-7 and triple-negative breast cancer cell line MDA-MB-231. The author concluded that OCT-Lcn2-Lipo also demonstrated in vitro anti-angiogenic effects in MCF-7 and MDA-MB-231 cells by reducing VEGF-A and reducing the endothelial cells (HUVEC) migration levels. This approach may be useful in inhibiting angiogenesis in

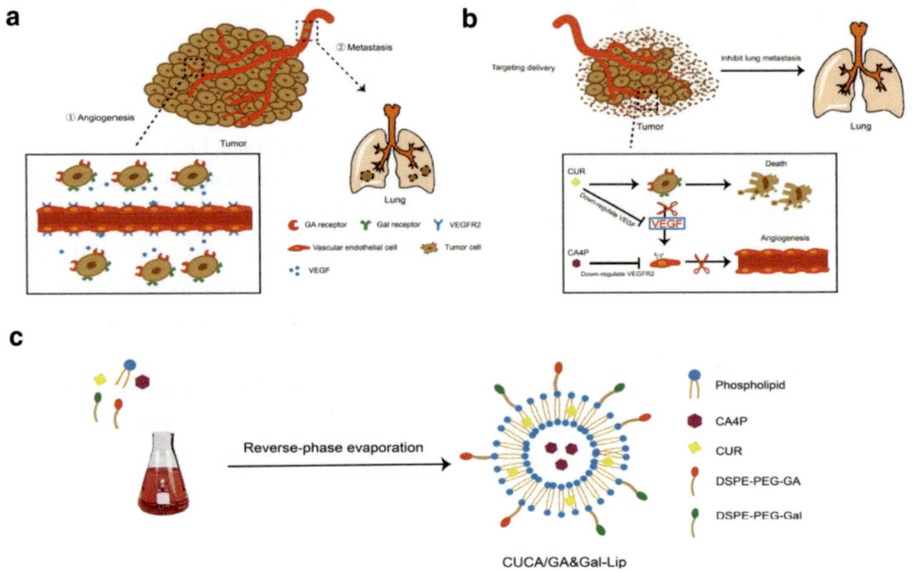

Fig. 4.1 **a** Schematic illustration of angiogenesis and lung metastasis. **b** Schematic illustration of anti-tumor mechanism and inhibit lung metastasis. **c** Illustration of liposomal preparations

MBC (Gote and Pal [39]). Similarly, several research groups developed anti-angiogenic drugs based on organic liposomes and studied their efficacies [40–45].

4.2 Solid Lipid Nanoparticles (SLN)

Sloid Lipid nanoparticles are the colloidal nanocarriers of a phospholipid monolayer, emulsifier, and water [46]. These are also known as zero-dimensional nanomaterials [47]. The lipid component may be triglycerides, waxes, fatty acids, steroids, and PEGylated lipids. Unlike traditional liposomes, SLNs have got a micelle-like structure surrounded by which the drug is trapped in a non-aqueous core. Examples include mitoxantrone- encaptsulated SLN, which has shown reduced toxicity and enhanced bioavailability [48]. Doxorubicin and idarubicin co-loaded SLNs showed positive results in P388/ADR leukemia cells and the murine leukemia mouse model [49]. Teucrium polium extract (TPE) is a natural product with potent anti-angiogenic and anti-cancer activity because of its terpenoid and flavonoid content [50]. For this, Roumi et al. 2021, synthesize solid lipid nanoparticles (SLN) containing T. polium extract (TPE-SLNs) to evaluate its anti-cancer effect. Formulations of TPE-SLNs were prepared using high-shear homogenization followed by the ultra-sonication technique [50]. According to the results of this study, solid lipid nanoparticles (SLN) containing T. polium extract showed their anti-cancer effects through various strategies such as inhibiting angiogenesis, inhibiting inflammation, and inducing apoptosis on Ntra-2 cells. These results make it possible to apply these nanocarriers in the loading and transport of drugs to treat various diseases such as cancer [50].

Further, Lo et al. 2020, developed liposomes and solid lipid nanoparticles (SLNs) altered with a pH-sensitive, self-destructive polyethylene glycol (PEG) shell and distinct peptides as irinotecan and miR-200 nanovectors to enhance the tumor-specific accumulation [51]. They found that the cleavable PEG layer responded to low extracellular pH and that the cell-penetrating peptide (CPP) and targeting peptides were exposed to improve the uptake and release of miR-200 and irinotecan into Head and Neck Cancer (HNC)-human tongue squamous carcinoma (SAS) cells [51]. Their findings also suggested chemotherapy and gene therapy co-treatment involving pH-sensitive and targeting peptide-modified nanoparticles may be an innovative strategy for HNC treatment [51]. Banerjee and his colleague (2016) constructed Tyr-3-octreotide (TOC)-modified solid lipid nanoparticles (SLN) loaded with paclitaxel (PTX) to enable tumor neovasculature and tumor cells dual-targeting chemotherapy, and the result indicate that PSM holds great potential in improving the efficacy of anti-glioma therapy [52].

4.3 Polymeric Based Nanoparticles

Polymeric nanoparticles (PNPs) are colloidal macromolecules with specific structural structural designs formed by several monomers [53]. The drug is either entrapped or connected to the NP's exterior, creating a nanocapsule to achieve controlled drug release in the target [54]. Primarily, PNPs were created up of non-biodegradable polymers such as polyacrylamide, polymethylmethacrylate (PMMA), and polystyrene [55]. However, these accumulations led to toxicity due to difficulty eliminating these from the system [55]. Biodegradable polymers like polylactic acid, poly(amino acids), alginate, chitosan, and albumin are now being used to reduce toxicity and improve drug release and biocompatibility [56]. Proven research has reflected coating PNPs with polysorbates and using the surfactant effect. Exterior coating enriches NPs' interactions with the endothelial cell membrane of the blood–brain barrier (BBB) [57, 58]. A study revealed that nanoparticles encaptsulated with indomethacin affected a significant reduction in the size of the tumor and superior survival in a xenograft glioma model in rats [59]. This is a developing field with more than ten polymeric NPs containing anti-cancer drugs under clinical training. A few examples include PEG-camptothecin (Prothecan), paclitaxel poliglumex (Xyotax), and Altered dextran-camptothecin (DE 310), HPMA copolymer-platinate (AP 5280), HPMA copolymer-DACH-platinate (AP5346), HPMA copolymer-paclitaxel (PNU166945), and HPMA copolymer-doxorubicin galactosamine (PK2) [60].

The use of nano-sized PLA particles has achieved much attention in drug delivery applications [60, 61]; one of the first reports on using these nanosized particles in an anti-angiogenic strategy was published by Burt et al. in 1995 [62, 63]. The use of nanoparticles containing copolymers made of PCL and other biocompatible polymers (e.g., PEG) is suggested to improve anti-angiogenic efficacy and, thereby, anti-cancer potential in vivo compared to PLA [64]. In contrast, it has been reported that electrospun PLA nanofibers could increase the proliferation of ECs in vitro [65]. Moreover, surface functionalization of PLA has been proposed as an approach to increase its pro-angiogenic properties; polyethyleneimine (PEI) and polyacrylic acid (PAC)-coated electrospun PLA nanofibers significantly promoted angiogenesis both in vitro and in vivo (Gigliobianco et al. 2015) [66]. PLA may be a suitable platform for delivering a range of pro-angiogenic molecules, such as VEGF [67].

There are few studies in the literature in which PCL was used for pro-and anti-angiogenic applications [68]. Jiang et al. could successfully prepare PCL nanofibers containing VEGF-encapsulated gelatin particles to enhance MSCs differentiation and angiogenesis of ECs [69]. PLGA is another FDA-approved pharmaceutical product, extensively used as a DDS in imaging diagnostics and therapy due to its favorable properties, such as biocompatibility and the controlled and sustained release of drugs [70]. In several studies, researchers have demonstrated the applicability of PLGA nanoparticles (pristine, chemically modified, or hybrids) to load and deliver anti-angiogenic molecules [9]. A tumor-vessel-recognizing and tumor-penetrating system were developed based on iNGR-modified PEG–PLGA nanoparticles to treat glioma in mice [71]. The modified nanoparticles could penetrate the tumor parenchyma and showed good cellular uptake in HUVECs, resulting in paclitaxel's enhanced anti-proliferative and anti-capillary tube formation activities in vitro [71]. Moreover, the results showed improved anti-angiogenic activity of the drug-loaded nano-carriers [71].

Focal adhesion kinase (FAK) and CD44 are predominantly involved in cancer development. Importantly, FAK has an active role in angiogenesis, cell apoptosis, and cell proliferation. However, heavily glycosylated CD44 has been involved in cancer metastasis. Hence, scientists developed PLGA nanoparticles by using short hairpin RNA (shRNA) against FAK and CD44 to inhibit tumor growth by inhibiting angiogenesis. These implants significantly reduced the number of new vessels in the pshFAK/PLGANP, pshCD44/PLGANP, and pshFAK-CD44/PLGANP compared with the control groups. Figure 4.2 shows that the most significant reduction was examined in the pshFAK-CD44/PLGANPs treated group compared to the single treatment, consistent with the CD31 staining results. In summary, the above results suggest that angiogenesis within the human ovarian cancer xenografts was inhibited in mice treated with the pshFAK/PLGANPs or pshCD44/PLGANPs, pshFAK-CD44/PLGANPs shRNA bearing nanoparticles. The effect was excellent in the pshFAK-CD44/PLGANPs group, indicating an advantage in targeting both proteins concurrently [72].

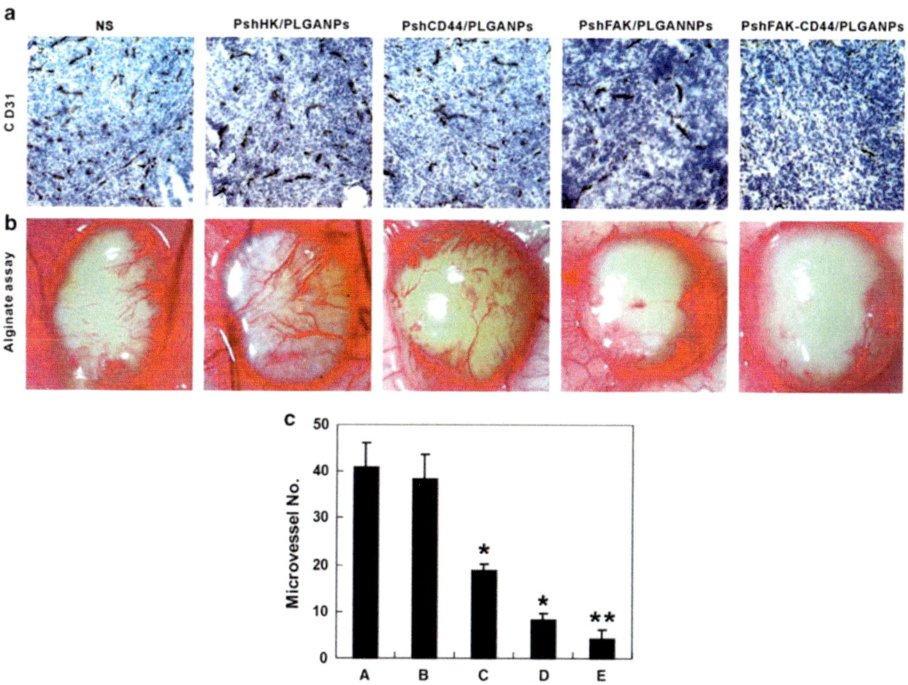

Fig. 4.2 Inhibition of angiogenesis by pshFAK-CD44/PLGANPs in vivo. **a** Representative images of immunohistochemical CD31 staining. **b** Vessel formation in an alginate bead assay. **c** Each group determined microvessel density (MVD) after immunohistochemical staining for CD31. (A) NS group (0.9% sodium chloride); (B) empty-vector (pshHK/PLGANPs) group; (C) pshCD44/PLGANPs group; (D) pshFAK/ PLGANPs group and (E) pshFAK-CD44/PLGANPs group. The results are presented as mean ± s.d. (n ¼ 5). *Po0.001 versus the other three groups; **Po0.001 versus the other four groups. FAK, focal adhesion kinase; PLGANPs, poly D, L-lactide-co-glycolide acid nanoparticles. NS, NaCl solution [72]

It is worth noting that numerous research groups have used PLGA nanoparticles to load pro-angiogenic biomolecules (e.g., aptamers) and other chemicals to improve angiogenesis and subsequently accelerate tissue healing [9]. As an FDA-approved product, chitosan in micro-and nanosized formulations is commonly used in many biomedical applications, from wound healing to drug delivery [70]. The biological activities of chitosan can be summarized as anti-microbial, anti-oxidant, and anti-cancer [73]. Furthermore, it has been reported that chitosan nanoparticles can inhibit angiogenesis in a dose- and time-dependent manner in cancer models in vivo [74]. The suppression of VEGFR-2 and subsequent blockage of VEGF is proposed to explain the anti-angiogenic activity of chitosan nanoparticles [75]. The anti-angiogenic activity was also observed in the case of depolymerized chitosan products, i.e., water-soluble low-molecular-weight chitosan (LMWC) and chito-oligosaccharides (COs) [73]. Furthermore, it should be stated that chitosan has also been used as a drug delivery system in pro-and anti-angiogenic applications [76].

4.4 Dendrimers

Dendrimers are spherical polymeric macromolecules with well-characterized hyperbranched construction. Highly branched structures are the characteristic aspect of dendrimers [77]. Characteristically, the production of dendrimers is established by reacting an ammonia core with acrylic acid [78, 79]. This reaction forms a tri-acid molecule that further reacts with ethylenediamine to yield tri-amine, a GO product [78]. This produce further reacts with acrylic acid to give rise to hexa-acid, producing hexa-amine (Generation 1) development [60]. Typically, the size of the dendrimers ranges from 1–10 nm. However, the size may reach up to 15 nm. Given their specific structure like defined molecular weight, adjustable branches, bioavailability, and charge, these are used to target nucleic acids [60]. Various dendrimers that are widely used are polyamidoamine (PAMAM), PPI (polypropylenimine), PEG (poly(ethyleneglycol)), and TEA (triethanolamine) (Lim et al. 2013) [79]. A PAMAM dendrimer was initially designed to achieve MDR management. DNA assembled PAMAM dendrimers have been described extensively. As compared with animals treated with single-agent chemotherapy, the synthesized dendrimers significantly delayed the growth of epithelial cancer xenografts [80]. Azimifar and his colleague (2021) introduced a novel delivery system based on poly-(amidoamine) (PAMAM) (G5) to deliver a plasmid encoding IL-12 to MSCs. Initially, 30% of the amine surface of PAMAM was substituted by 10-bromodecanoic acid. Then, the low molecular weight of the protamine peptide was conjugated to PAMAM and PAMAM-alkyl with N-succinimidyl 3-(2-pyridyldithio) propionate as a linker [81]. The results showed that alkyl-peptide modified PAMAM with low toxicity had a higher potential to deliver green fluorescent protein and IL-12 genes to stem cells than PMAMAM, PAMAM-alkyl, and PAMAM-peptide [81]. These engineered stem cells had a more extraordinary ability to migrate to cancer cells than normal cells. It can be concluded that engineered stem cells containing the IL-12 gene can be considered an efficient cell carrier for cancer immunotherapy. Further clinical studies are needed to confirm these results [81]. Anti-microbial polycationic dendriers G3KL and G3RL, able to kill multidrug resistant *P. aeruginosa,* have been previously developed by Abdel-Sayed et al. [82]. Combining these dendrimers with a class of biological bandages made of progenitor skin cells, which secrete growth factors, could positively impact wound-healing processes [82]. However, polycations are known to be used as anti-angiogenic agents for tumor suppression. The results show that anti-microbial dendrimers can be used in combination with biological bandages and could potentially improve the healing process with enhanced angiogenesis [82].

4.5 Nanoemulsions

Nanoemulsions are colloidal NPs with diverse combinations of an oil droplet in aqueous media ranging from 10–1000 nm [83]. Three representative types of nanoemulsions can be made in (1) oil-in-water system, (2) water-in-oil system, and (3) bi-continuous nanoemulsions. Membrane-modified nanoemulsions have been extensively studied [84]. For instance, nanoemulsions load up with spirulina and paclitaxel demonstrated an improved anti-tumor effect by regulating immunity [85]. Nanoemulsion consisting of temozolomide, rapamycin, and bevacizumab is known to treat advanced melanoma [86]. Nanoemulsions are different from liposomes and certainly have improved characteristics over others, such as optical clarity, stability, and biodegradability [87]. However, there are tasks to clinical applications of these nanoemulsions as these involve high temperature and pressure and instruments such as homogenizers and microfluidizers that are expensive [88]. Azani et al. 2021, produced a nanoemulsion-based drug delivery system containing *Ferula assa-foetida* (FA) essential oil (FAEO) and evaluated its anti-oxidant and anti-cancer activity on both sides of MCF-7 cells and murine mammary cancer tissue and characterized its anti-tumor impact was evaluated on murine mammary tumor models [89]. The FAEO-NE exhibited a meaningful anti-oxidant activity. Also, its significant cell-selective cytotoxic, apoptotic, and anti-angiogenic impacts on MCF-7 cancer cells indicated its anti-cancer potential [89]. An another group of the researcher designed and optimization of nanoemulsion (NE) gel formulation for the concurrent delivery of Curcumin, resveratrol, and thymoquinone, and the in vivo studies in the Balb/c mice model displayed improved anti-psoriatic conditions which designated that the triple natural bio-actives mixture in nanoemulgel preparation is effective in the management of psoriasis [90]. *Linum usitatissimum* seed essential oil (LSEO) has been used to reduce the risk of prostate and colon cancer [91]. For this, Keykhasalar and his researcher group developed and optimized the bio-accessibility and bio-compatibility of LSEO to evaluate its cytotoxic, apoptotic, and anti-angiogenic impact on the human ovarian cancer cell line A2780 [91]. The author concluded that the significant dose-dependent cytotoxic impact of LSEO-NE was detected in the A2780 cells and not in HFF cells [91]. The apoptotic gene expression profile confirmed the A2780 cell apoptosis death. Moreover, the reduction in length and number of blood vessels in the CAM assay demonstrated the anti-angiogenic activity of LSEO-NE [91]. *Citrus lemon* essential oil (CLEO) has been announced as an anti-oxidant solid mixture. However, it is not effective enough due to its hydrophobic nature [92]. Nanoemulsions develop the drugs bio-compatibility in aqueous conditions. In the same way, Rad and group developed CLEO-NE droplets that drastically increased Cas-3 in A549 cells and reduced angiogenesis in chick embryo chorioallantoic membrane [92]. Li et al. developed a pulmonary delivery (p.d.) system based on perfluorocarbon (PFC) nanoemulsions for combined delivery of a partially fluorinated polymeric CXCR4 antagonist (FM) and anti-STAT3 small interfering RNA (siRNA). They suggested that the nanoemulsions for p.d. of combination anti-cancer treatments and as a promising procedure to treat lung metastasis [93].

4.6 Polymeric Nanofibers

Polymeric nanofibers are among the most widely-applied constructs in biomedicine, from anti-tumor strategies to tissue healing [94]. Nanofibers exhibit some attractive properties, including large specific surface area, controllable pore size, and tunable drug release profiles, making them highly-promising candidates for anti-cancer applications [95]. Recently, Abid et al. 2019, reviewed the anti-cancer applications of electrospun polymeric nanofibers loaded with various chemicals and drugs, including doxorubicin, paclitaxel, and curcumin [96]. Apart from electrospun nanofibers, several studies showed the utility of synthetic nanofibrous peptide scaffolds to mimic small molecules' pro-angiogenic and anti-angiogenic activity, including heparin and maspin [97, 98] For instance, Fan et al. investigated docetaxel- and curcumin-loaded nanofibrous microspheres made of PLA–PEO–PPO–PEO–PLA polymers as an injectable and sustained-release system for enhancing anti-colon cancer activity [99]. The results of the combined nanofibrous microsphere treatment showed a significant increase in the inhibition of angiogenesis and subsequent inhibition of colon cancer in mice [99]. On the contrary, several publications in which pro-angiogenic cargos were delivered using polymeric nanofibrous scaffolds produced by electrospinning and self-assembly procedures [65, 100]. Most pro-angiogenic nanofibers have been applied to accelerate tissue repair and regeneration, especially to promote wound healing [101].

4.7 Cyclodextrin Nanosponges

Cyclodextrins are commonly used as stabilisers to increase the drug consignment capacity of NPs [102]. Nanosponges are tiny, mesh-like structures [103]. Β-cyclodextrin nanosponges loaded with paclitaxel have shown sound cytotoxic effects in MCF-7 cell line culture [60]. Similarly, camptothecin has demonstrated enhanced solubility and stability when formulated with cyclodextrin-based nanosponges [104]. Camptothecin (CPT), a pentacyclic alkaloid, is an inhibitor of DNA Topoisomerase-I and shows a broad spectrum of anti-cancer activities. CPT use has been hampered by poor aqueous solubility and a high degradation rate [105]. β-cyclodextrin-nanosponges (CN-CPT) overcomes these disadvantages and enhances the CPT's inhibitory impact on DU145 prostate tumor cell lines and PC-3 growth in vitro [105]. This research extends these observations by showing that CN-CPT significantly inhibits the adhesion and migration of these tumor cells and their STAT3 phosphorylation and suggests that β-cyclodextrin nanosponge nanotechnology is a potential nanocarrier for the delivery of anti-cancer drugs in the treatment of prostate cancers [105].

Organic nanoparticles are up-and-coming candidates for the growth of drug delivery systems for cancer therapy, and their success was already determined in clinical applications. Nanoparticles-based drug delivery systems are superior to conventional anticancer drugs because they can decrease the systemic side effects that patients must endure under traditional chemotherapy by ensuring that the cytotoxic levels of the drugs are only present at the tumor sites. The future perspectives in cancer treatment refer to obtaining a multifunctional system able to apply for simultaneous therapy of cancers and in vivo imaging (CT or MRI contrast agents) by using angiogenesis molecular marker targets. These nanoparticles able to carry: one or more drugs, a special targeting moiety, a cell-penetrating agent, a stabilizing polymer for biocompatibility, a stimulus-sensitive element for limited release of drugs, and an imaging agent. Several types of nanoparticles, including polymeric nanoparticles, polymeric micelles, dendrimers, and quantum dots, have been evaluated for their suitability as multifunctional nanoparticles. The researchers are trying to locate the best administration route (intravenous, intra-tumoral, and intra-peritoneal) for these delivery systems.

References

1. K.I. Alcaraz, T.L. Wiedt, E.C. Daniels, K.R. Yabroff, C.E. Guerra, R.C. Wender, Understanding and addressing social determinants to advance cancer health equity in the United States: a blueprint for practice, research, and policy, CA: a cancer journal for clinicians 70(1) (2020) 31–46.
2. M. Najafi, J. Majidpoor, H. Toolee, K. Mortezaee, The current knowledge concerning solid cancer and therapy, Journal of biochemical and molecular toxicology 35(11) (2021) e22900.
3. G. Bor, I.D. Mat Azmi, A. Yaghmur, Nanomedicines for cancer therapy: Current status, challenges and future prospects, Therapeutic delivery 10(2) (2019) 113–132.
4. S. Saravanan, S. Vimalraj, K. Pavani, R. Nikarika, V.N. Sumantran, Intussusceptive angiogenesis as a key therapeutic target for cancer therapy, Life sciences 252 (2020) 117670.
5. F.E. Pulous, J.C. Carnevale, Z. Al-Yafeai, B.H. Pearson, J.A. Hamilton, C.J. Henry, A.W. Orr, B.G. Petrich, Talin-dependent integrin activation is required for endothelial proliferation and postnatal angiogenesis, Angiogenesis 24(1) (2021) 177-190.
6. D.A. Chistiakov, A.N. Orekhov, Y.V. Bobryshev, The role of miR-126 in embryonic angiogenesis, adult vascular homeostasis, and vascular repair and its alterations in atherosclerotic disease, Journal of molecular and cellular cardiology 97 (2016) 47–55.
7. L. Morbidelli, E. Terzuoli, S. Donnini, Use of nutraceuticals in angiogenesis-dependent disorders, Molecules 23(10) (2018) 2676.
8. M. Szczuko, J. Palma, J. Kikut, N. Komorniak, M. Ziętek, Changes of lipoxin levels during pregnancy and the monthly-cycle, condition the normal course of pregnancy or pathology, Inflammation Research 69(9) (2020) 869–881.
9. S. Kargozar, F. Baino, S. Hamzehlou, M.R. Hamblin, M. Mozafari, Nanotechnology for angiogenesis: opportunities and challenges, Chemical Society Reviews 49(14) (2020) 5008–5057.
10. S. Jain, K. Raza, A.K. Agrawal, A. Vaidya, Nanotechnology Applications for Cancer Chemotherapy, Elsevier2020.
11. J. Jászai, M.H. Schmidt, Trends and challenges in tumor anti-angiogenic therapies, Cells 8(9) (2019) 1102.

12. X. Chang, A. Lochner, H.-H. Wang, S. Wang, H. Zhu, J. Ren, H. Zhou, Coronary microvascular injury in myocardial infarction: perception and knowledge for mitochondrial quality control, Theranostics 11(14) (2021) 6766.
13. K.M. Cook, W.D. Figg, Angiogenesis inhibitors: current strategies and future prospects, CA: a cancer journal for clinicians 60(4) (2010) 222–243.
14. G. Gasparini, R. Longo, M. Toi, N. Ferrara, Angiogenic inhibitors: a new therapeutic strategy in oncology, Nature clinical practice Oncology 2(11) (2005) 562–577.
15. A.S. Oguntade, F. Al-Amodi, A. Alrumayh, M. Alobaida, M. Bwalya, Anti-angiogenesis in cancer therapeutics: The magic bullet, Journal of the Egyptian National Cancer Institute 33(1) (2021) 1–11.
16. P. Bhattarai, S. Hameed, Z. Dai, Recent advances in anti-angiogenic nanomedicines for cancer therapy, Nanoscale 10(12) (2018) 5393–5423.
17. K.J. Kim, B. Li, J. Winer, M. Armanini, N. Gillett, H.S. Phillips, N. Ferrara, Inhibition of vascular endothelial growth factor-induced angiogenesis suppresses tumour growth in vivo, Nature 362(6423) (1993) 841–844.
18. A. Nasir, Angiogenic signaling pathways and anti-angiogenic therapies in human cancer, Predictive biomarkers in oncology, Springer2019, pp. 243–262.
19. R.P. Million, Therapeutic area crossroads: anti-angiogenesis, Nature Reviews. Drug Discovery 7(2) (2008) 115.
20. K. Zirlik, J. Duyster, Anti-angiogenics: current situation and future perspectives, Oncology research and treatment 41(4) (2018) 166–171.
21. K.B. Neves, A.C. Montezano, N.N. Lang, R.M. Touyz, Vascular toxicity associated with anti-angiogenic drugs, Clinical Science 134(18) (2020) 2503–2520.
22. J. Jeevanandam, A. Barhoum, Y.S. Chan, A. Dufresne, M.K. Danquah, Review on nanoparticles and nanostructured materials: history, sources, toxicity and regulations, Beilstein journal of nanotechnology 9(1) (2018) 1050–1074.
23. S. Siddique, J.C. Chow, Application of nanomaterials in biomedical imaging and cancer therapy, Nanomaterials 10(9) (2020) 1700.
24. M.J. Evers, S.I. van de Wakker, E.M. de Groot, O.G. de Jong, J.J. Gitz-François, C.S. Seinen, J.P. Sluijter, R.M. Schiffelers, P. Vader, Functional siRNA Delivery by Extracellular Vesicle–Liposome Hybrid Nanoparticles, Advanced Healthcare Materials 11(5) (2022) 2101202.
25. K.P. Sathe, V.B. Bangar, Liposomes: An Overview, (2021).
26. H. Mishra, V. Chauhan, K. Kumar, D. Teotia, A comprehensive review on Liposomes: a novel drug delivery system, Journal of Drug Delivery and Therapeutics 8(6) (2018) 400–404.
27. T.O. Olusanya, R.R. Haj Ahmad, D.M. Ibegbu, J.R. Smith, A.A. Elkordy, Liposomal drug delivery systems and anticancer drugs, Molecules 23(4) (2018) 907.
28. N. Maghsoudnia, R.B. Eftekhari, A.N. Sohi, A. Zamzami, F.A. Dorkoosh, Application of nano-based systems for drug delivery and targeting: a review, Journal of Nanoparticle Research 22(8) (2020) 1–41.
29. E. Beltrán-Gracia, A. López-Camacho, I. Higuera-Ciapara, J.B. Velázquez-Fernández, A.A. Vallejo-Cardona, Nanomedicine review: Clinical developments in liposomal applications, Cancer Nanotechnology 10(1) (2019) 1–40.
30. H. Sun, X. Guo, S. Zeng, Y. Wang, J. Hou, D. Yang, S. Zhou, A multifunctional liposomal nanoplatform co-delivering hydrophobic and hydrophilic doxorubicin for complete eradication of xenografted tumors, Nanoscale 11(38) (2019) 17759–17772.
31. T.M. Allen, P.R. Cullis, Liposomal drug delivery systems: from concept to clinical applications, Advanced drug delivery reviews 65(1) (2013) 36–48.
32. R.-D. Hofheinz, S.U. Gnad-Vogt, U. Beyer, A. Hochhaus, Liposomal encapsulated anti-cancer drugs, Anti-cancer drugs 16(7) (2005) 691–707.

33. S. Gavas, S. Quazi, T.M. Karpiński, Nanoparticles for cancer therapy: current progress and challenges, Nanoscale Research Letters 16(1) (2021) 1–21.
34. V.-F. Rauca, L. Patras, L. Luput, E. Licarete, V.-A. Toma, A. Porfire, A.C. Mot, E. Rakosy-Tican, A. Sesarman, M. Banciu, Remodeling tumor microenvironment by liposomal codelivery of DMXAA and simvastatin inhibits malignant melanoma progression, Scientific reports 11(1) (2021) 1–15.
35. M. Mashreghi, M. Faal Maleki, M. Karimi, F. Kalalinia, A. Badiee, M.R. Jaafari, Improving anti-tumour efficacy of PEGylated liposomal doxorubicin by dual targeting of tumour cells and tumour endothelial cells using anti-p32 CGKRK peptide, Journal of drug targeting 29(6) (2021) 617–630.
36. M. Renault-Mahieux, V. Vieillard, J. Seguin, P. Espeau, D.T. Le, R. Lai-Kuen, N. Mignet, M. Paul, K. Andrieux, Co-Encapsulation of Fisetin and Cisplatin into Liposomes for Glioma Therapy: From Formulation to Cell Evaluation, Pharmaceutics 13(7) (2021) 970.
37. M. Petrini, W.J. Lokerse, A. Mach, M. Hossann, O.M. Merkel, L.H. Lindner, Effects of Surface Charge, PEGylation and Functionalization with Dipalmitoylphosphatidyldiglycerol on Liposome–Cell Interactions and Local Drug Delivery to Solid Tumors via Thermosensitive Liposomes, International Journal of Nanomedicine 16 (2021) 4045.
38. F. Wang, Y. Li, H. Jiang, C. Li, Z. Li, C. Qi, Z. Li, Z. Gao, B. Zhang, J. Wu, Dual-Ligand-Modified Liposomes Co-Loaded with Anti-Angiogenic and Chemotherapeutic Drugs for Inhibiting Tumor Angiogenesis and Metastasis, International Journal of Nanomedicine 16 (2021) 4001.
39. V. Gote, D. Pal, Octreotide-Targeted Lcn2 siRNA PEGylated Liposomes as a Treatment for Metastatic Breast Cancer, Bioengineering (Basel, Switzerland) 8(4) (2021).
40. A. Sesarman, L. Tefas, B. Sylvester, E. Licarete, V. Rauca, L. Luput, L. Patras, M. Banciu, A. Porfire, Anti-angiogenic and anti inflammatory effects of long-circulating liposomes co-encapsulating curcumin and doxorubicin on C26 murine colon cancer cells, Pharmacological reports 70(2) (2018) 331–339.
41. A.K. Rajora, D. Ravishankar, H. Zhang, J.M. Rosenholm, Recent advances and impact of chemotherapeutic and antiangiogenic nanoformulations for combination cancer therapy, Pharmaceutics 12(6) (2020) 592.
42. A.M. Ayoub, M.U. Amin, G. Ambreen, A.A. Dayyih, A.M. Abdelsalam, A. Somaida, K. Engelhardt, M. Wojcik, J. Schäfer, U. Bakowsky, Photodynamic and antiangiogenic activities of parietin liposomes in triple negative breast cancer, Materials Science and Engineering: C (2021) 112543.
43. A. Mukherjee, V.S. Madamsetty, M.K. Paul, S. Mukherjee, Recent Advancements of Nanomedicine towards Antiangiogenic Therapy in Cancer, Int J Mol Sci 21(2) (2020) 455.
44. D.K. Chang, C.Y. Chiu, S.Y. Kuo, W.C. Lin, A. Lo, Y.P. Wang, P.C. Li, H.C. Wu, Antiangiogenic targeting liposomes increase therapeutic efficacy for solid tumors, The Journal of biological chemistry 284(19) (2009) 12905–16.
45. L.W. Allahou, S.Y. Madani, A. Seifalian, Investigating the Application of Liposomes as Drug Delivery Systems for the Diagnosis and Treatment of Cancer, International Journal of Biomaterials 2021 (2021) 3041969.
46. V. Mishra, K.K. Bansal, A. Verma, N. Yadav, S. Thakur, K. Sudhakar, J.M. Rosenholm, Solid Lipid Nanoparticles: Emerging Colloidal Nano Drug Delivery Systems, Pharmaceutics 10(4) (2018) 191.
47. A. Puri, K. Loomis, B. Smith, J.-H. Lee, A. Yavlovich, E. Heldman, R. Blumenthal, Lipid-based nanoparticles as pharmaceutical drug carriers: from concepts to clinic, Crit Rev Ther Drug Carrier Syst 26(6) (2009) 523–580.

48. A. Granja, R. Lima-Sousa, C.G. Alves, D. de Melo-Diogo, M. Pinheiro, C.T. Sousa, I.J. Correia, S. Reis, Mitoxantrone-loaded lipid nanoparticles for breast cancer therapy–Quality-by-design approach and efficacy assessment in 2D and 3D in vitro cancer models, International Journal of Pharmaceutics 607 (2021) 121044.
49. P. Ma, X. Dong, C.L. Swadley, A. Gupte, M. Leggas, H.C. Ledebur, R.J. Mumper, Development of idarubicin and doxorubicin solid lipid nanoparticles to overcome Pgp-mediated multiple drug resistance in leukemia, J Biomed Nanotechnol 5(2) (2009) 151–161.
50. S. Roumi, M.H. Tabrizi, A. Eshaghi, N. Abbasi, Teucrium polium extract-loaded solid lipid nanoparticles: A design and in vitro anticancer study, Journal of food biochemistry 45(9) (2021) e13868.
51. Y.-L. Lo, C.-H. Chang, C.-S. Wang, M.-H. Yang, A.M.-Y. Lin, C.-J. Hong, W.-H. Tseng, PEG-coated nanoparticles detachable in acidic microenvironments for the tumor-directed delivery of chemo- and gene therapies for head and neck cancer, Theranostics 10(15) (2020) 6695–6714.
52. I. Banerjee, K. De, D. Mukherjee, G. Dey, S. Chattopadhyay, M. Mukherjee, M. Mandal, A.K. Bandyopadhyay, A. Gupta, S. Ganguly, M. Misra, Paclitaxel-loaded solid lipid nanoparticles modified with Tyr-3-octreotide for enhanced anti-angiogenic and anti-glioma therapy, Acta biomaterialia 38 (2016) 69–81.
53. L.H. Madkour, Chapter 14 - Polymer nanoparticle drug-nucleic acid combinations, in: L.H. Madkour (Ed.), Nucleic Acids as Gene Anticancer Drug Delivery Therapy, Academic Press2019, pp. 241–255.
54. I. Ali, M. Alsehli, L. Scotti, M. Tullius Scotti, S.-T. Tsai, R.-S. Yu, M.F. Hsieh, J.-C. Chen, Progress in Polymeric Nano-Medicines for Theranostic Cancer Treatment, Polymers (Basel) 12(3) (2020) 598.
55. H. Idrees, S.Z.J. Zaidi, A. Sabir, R.U. Khan, X. Zhang, S.-U. Hassan, A Review of Biodegradable Natural Polymer-Based Nanoparticles for Drug Delivery Applications, Nanomaterials (Basel) 10(10) (2020) 1970.
56. R. Song, M. Murphy, C. Li, K. Ting, C. Soo, Z. Zheng, Current development of biodegradable polymeric materials for biomedical applications, Drug Des Devel Ther 12 (2018) 3117–3145.
57. N. Hartl, F. Adams, O.M. Merkel, From adsorption to covalent bonding: Apolipoprotein E functionalization of polymeric nanoparticles for drug delivery across the blood-brain barrier, Adv Ther (Weinh) 4(1) (2020) 2000092.
58. M. Elsabahy, K.L. Wooley, Design of polymeric nanoparticles for biomedical delivery applications, Chemical Society Reviews 41(7) (2012) 2545–2561.
59. A. Bernardi, E. Braganhol, E. Jäger, F. Figueiró, M.I. Edelweiss, A.R. Pohlmann, S.S. Guterres, A.M. Battastini, Indomethacin-loaded nanocapsules treatment reduces in vivo glioblastoma growth in a rat glioma model, Cancer letters 281(1) (2009) 53–63.
60. S. Gavas, S. Quazi, T.M. Karpiński, Nanoparticles for Cancer Therapy: Current Progress and Challenges, Nanoscale research letters 16(1) (2021) 173–173.
61. S. Yadav, A.K. Sharma, P. Kumar, Nanoscale Self-Assembly for Therapeutic Delivery, Frontiers in Bioengineering and Biotechnology 8 (2020).
62. K.H. Bae, H.J. Chung, T.G. Park, Nanomaterials for cancer therapy and imaging, Mol Cells 31(4) (2011) 295–302.
63. R.S. Kerbel, Tumor angiogenesis: past, present and the near future, Carcinogenesis 21(3) (2000) 505–515.
64. J. Karlsson, H.J. Vaughan, J.J. Green, Biodegradable Polymeric Nanoparticles for Therapeutic Cancer Treatments, Annu Rev Chem Biomol Eng 9 (2018) 105–127.
65. S. Nazarnezhad, F. Baino, H.-W. Kim, T.J. Webster, S. Kargozar, Electrospun Nanofibers for Improved Angiogenesis: Promises for Tissue Engineering Applications, Nanomaterials (Basel) 10(8) (2020) 1609.

66. S.N.A. Bukhari, Emerging Nanotherapeutic Approaches to Overcome Drug Resistance in Cancers with Update on Clinical Trials, Pharmaceutics 14(4) (2022) 866.
67. N.S. Bhise, R.B. Shmueli, J.C. Sunshine, S.Y. Tzeng, J.J. Green, Drug delivery strategies for therapeutic angiogenesis and antiangiogenesis, Expert Opin Drug Deliv 8(4) (2011) 485–504.
68. Z. Mohammadalizadeh, E. Bahremandi-Toloue, S. Karbasi, Recent advances in modification strategies of pre- and post-electrospinning of nanofiber scaffolds in tissue engineering, Reactive and Functional Polymers 172 (2022) 105202.
69. Y.-C. Jiang, X.-F. Wang, Y.-Y. Xu, Y.-H. Qiao, X. Guo, D.-F. Wang, Q. Li, L.-S. Turng, Polycaprolactone Nanofibers Containing Vascular Endothelial Growth Factor-Encapsulated Gelatin Particles Enhance Mesenchymal Stem Cell Differentiation and Angiogenesis of Endothelial Cells, Biomacromolecules 19(9) (2018) 3747–3753.
70. M. Mir, N. Ahmed, A.u. Rehman, Recent applications of PLGA based nanostructures in drug delivery, Colloids and Surfaces B: Biointerfaces 159 (2017) 217–231.
71. T. Kang, X. Gao, Q. Hu, D. Jiang, X. Feng, X. Zhang, Q. Song, L. Yao, M. Huang, X. Jiang, Z. Pang, H. Chen, J. Chen, iNGR-modified PEG-PLGA nanoparticles that recognize tumor vasculature and penetrate gliomas, Biomaterials 35(14) (2014) 4319–4332.
72. L. Zou, X. Song, T. Yi, S. Li, H. Deng, X. Chen, Z. Li, Y. Bai, Q. Zhong, Y. Wei, X. Zhao, Administration of PLGA nanoparticles carrying shRNA against focal adhesion kinase and CD44 results in enhanced antitumor effects against ovarian cancer, Cancer Gene Therapy 20(4) (2013) 242–250.
73. I. Aranaz, A.R. Alcántara, M.C. Civera, C. Arias, B. Elorza, A. Heras Caballero, N. Acosta, Chitosan: An Overview of Its Properties and Applications, Polymers (Basel) 13(19) (2021) 3256.
74. Y. Xu, Z. Wen, Z. Xu, Chitosan Nanoparticles Inhibit the Growth of Human Hepatocellular Carcinoma Xenografts through an Antiangiogenic Mechanism, Anticancer Research 29(12) (2009) 5103.
75. Y. Li, W. Wang, Y. Zhang, X. Wang, X. Gao, Z. Yuan, Y. Li, Chitosan sulfate inhibits angiogenesis via blocking the VEGF/VEGFR2 pathway and suppresses tumor growth in vivo, Biomaterials science 7(4) (2019) 1584–1597.
76. S. Mukherjee, C.R. Patra, Therapeutic application of anti-angiogenic nanomaterials in cancers, Nanoscale 8(25) (2016) 12444–12470.
77. S. Singh, G. Singh, S. Sehrawat, P. Rawat, N. Molugulu, V. Singh, F.J. Ahmed, P. Kesharwani, Chapter 25-Conclusion and future considerations of dendrimers, in: P. Kesharwani (Ed.), Dendrimer-Based Nanotherapeutics, Academic Press2021, pp. 449–458.
78. C. Qu, W. Shu, F. Xie, Y. Liu, R. Li, C. Pei, J. Wan, Dendrimer-Modified Silica Nanoparticles for Efficient Enrichment of Low-Concentration Peptides, Applied Biochemistry and Biotechnology (2022).
79. F. Najafi, M. Salami-Kalajahi, H. Roghani-Mamaqani, A review on synthesis and applications of dendrimers, Journal of the Iranian Chemical Society 18(3) (2021) 503–517.
80. A.K. Sharma, A. Gothwal, P. Kesharwani, H. Alsaab, A.K. Iyer, U. Gupta, Dendrimer nanoarchitectures for cancer diagnosis and anticancer drug delivery, Drug Discovery Today 22(2) (2017) 314–326.
81. M.A. Azimifar, Z. Salmasi, A. Doosti, N. Babaei, M. Hashemi, Evaluation of the efficiency of modified PAMAM dendrimer with low molecular weight protamine peptide to deliver IL-12 plasmid into stem cells as cancer therapy vehicles, Biotechnology Progress 37(4) (2021) e3175.

82. P. Abdel-Sayed, A. Kaeppeli, T. Siriwardena, T. Darbre, K. Perron, P. Jafari, J.-L. Reymond, D.P. Pioletti, L.A. Applegate, Anti-Microbial Dendrimers against Multidrug-Resistant P. aeruginosa Enhance the Angiogenic Effect of Biological Burn-wound Bandages, Scientific Reports 6(1) (2016) 22020.
83. E. Sánchez-López, M. Guerra, J. Dias-Ferreira, A. Lopez-Machado, M. Ettcheto, A. Cano, M. Espina, A. Camins, M.L. Garcia, E.B. Souto, Current Applications of Nanoemulsions in Cancer Therapeutics, Nanomaterials 9(6) (2019) 821.
84. P. Bharmoria, M. Bisht, M.C. Gomes, M. Martins, M.C. Neves, J.F. Mano, I. Bdikin, J.A.P. Coutinho, S.P.M. Ventura, Protein-olive oil-in-water nanoemulsions as encapsulation materials for curcumin acting as anticancer agent towards MDA-MB-231 cells, Scientific Reports 11(1) (2021) 9099.
85. M. Du, Z. Yang, W. Lu, B. Wang, Q. Wang, Z. Chen, L. Chen, S. Han, T. Cai, Y. Cai, Design and development of spirulina polysaccharide-loaded nanoemulsions with improved the antitumor effects of paclitaxel, Journal of microencapsulation 37(6) (2020) 403–412.
86. R. Cassano, M. Cuconato, G. Calviello, S. Serini, S. Trombino, Recent Advances in Nanotechnology for the Treatment of Melanoma, Molecules 26(4) (2021) 785.
87. S.M. Alshahrani, A judicious review on the applications of chemotherapeutic loaded nanoemulsions in cancer management, Journal of Drug Delivery Science and Technology 68 (2022) 103085.
88. E. Sánchez-López, M. Guerra, J. Dias-Ferreira, A. Lopez-Machado, M. Ettcheto, A. Cano, M. Espina, A. Camins, M.L. Garcia, E.B. Souto, Current Applications of Nanoemulsions in Cancer Therapeutics, Nanomaterials (Basel) 9(6) (2019) 821.
89. H. Azani, M. Homayouni Tabrizi, A. Neamati, F. Khadem, N. Khatamian, The Ferula Assa-foetida Essential Oil Nanoemulsion (FAEO-NE) as the Selective, Apoptotic, and Anti-Angiogenic Anticancer Compound in Human MCF-7 Breast Cancer Cells and Murine Mammary Tumor Models, Nutrition and cancer (2021) 1–11.
90. K. Khatoon, A. Ali, F.J. Ahmad, Z. Hafeez, M.M.A. Rizvi, S. Akhter, S. Beg, Novel nanoemulsion gel containing triple natural bio-actives combination of curcumin, thymoquinone, and resveratrol improves psoriasis therapy: in vitro and in vivo studies, Drug delivery and translational research 11(3) (2021) 1245–1260.
91. R. Keykhasalar, M.H. Tabrizi, P. Ardalan, N. Khatamian, The Apoptotic, Cytotoxic, and Antiangiogenic Impact of Linum usitatissimum Seed Essential Oil Nanoemulsions on the Human Ovarian Cancer Cell Line A2780, Nutrition and cancer 73(11-12) (2021) 2388–2396.
92. E. Yousefian Rad, M. Homayouni Tabrizi, P. Ardalan, S.M.R. Seyedi, S. Yadamani, P. Zamani-Esmati, N. Haghani Sereshkeh, Citrus lemon essential oil nanoemulsion (CLEO-NE), a safe cell-depended apoptosis inducer in human A549 lung cancer cells with anti-angiogenic activity, Journal of microencapsulation 37(5) (2020) 394–402.
93. Z. Li, G. Chen, L. Ding, Y. Wang, C. Zhu, K. Wang, J. Li, M. Sun, D. Oupicky, Increased Survival by Pulmonary Treatment of Established Lung Metastases with Dual STAT3/CXCR4 Inhibition by siRNA Nanoemulsions, Molecular therapy : the journal of the American Society of Gene Therapy 27(12) (2019) 2100–2110.
94. Z. Liu, S. Ramakrishna, X. Liu, Electrospinning and emerging healthcare and medicine possibilities, APL Bioeng 4(3) (2020) 030901–030901.
95. A. Luraghi, F. Peri, L. Moroni, Electrospinning for drug delivery applications: A review, Journal of Controlled Release 334 (2021) 463–484.
96. S. Abid, T. Hussain, Z.A. Raza, A. Nazir, Current applications of electrospun polymeric nanofibers in cancer therapy, Materials science & engineering. C, Materials for biological applications 97 (2019) 966–977.

97. X. Gao, X. Zhang, J. Song, X. Xu, A. Xu, M. Wang, B. Xie, E. Huang, F. Deng, S. Wei, Osteoinductive peptide-functionalized nanofibers with highly ordered structure as biomimetic scaffolds for bone tissue engineering, International journal of nanomedicine 10 (2015) 7109–7128.
98. R.H. Zha, S. Sur, J. Boekhoven, H.Y. Shi, M. Zhang, S.I. Stupp, Supramolecular assembly of multifunctional maspin-mimetic nanostructures as a potent peptide-based angiogenesis inhibitor, Acta biomaterialia 12 (2015) 1–10.
99. R. Fan, X. Li, J. Deng, X. Gao, L. Zhou, Y. Zheng, A. Tong, X. Zhang, C. You, G. Guo, Dual Drug Loaded Biodegradable Nanofibrous Microsphere for Improving Anti-Colon Cancer Activity, Scientific Reports 6(1) (2016) 28373.
100. S. Nemati, S.-j. Kim, Y.M. Shin, H. Shin, Current progress in application of polymeric nanofibers to tissue engineering, Nano Convergence 6(1) (2019) 36.
101. A.P. Veith, K. Henderson, A. Spencer, A.D. Sligar, A.B. Baker, Therapeutic strategies for enhancing angiogenesis in wound healing, Advanced drug delivery reviews 146 (2019) 97–125.
102. D.D. Gadade, S.S. Pekamwar, Cyclodextrin Based Nanoparticles for Drug Delivery and Theranostics, Adv Pharm Bull 10(2) (2020) 166–183.
103. P. Pandey, D. Purohit, H. Dureja, Nanosponges -A Promising Novel Drug Delivery System, Recent patents on nanotechnology 12(3) (2018) 180–191.
104. S. Swaminathan, L. Pastero, L. Serpe, F. Trotta, P. Vavia, D. Aquilano, M. Trotta, G. Zara, R. Cavalli, Cyclodextrin-based nanosponges encapsulating camptothecin: physicochemical characterization, stability and cytotoxicity, European journal of pharmaceutics and biopharmaceutics : official journal of Arbeitsgemeinschaft fur Pharmazeutische Verfahrenstechnik e.V 74(2) (2010) 193–201.
105. C.L. Gigliotti, R. Minelli, R. Cavalli, S. Occhipinti, G. Barrera, S. Pizzimenti, G. Cappellano, E. Boggio, L. Conti, R. Fantozzi, M. Giovarelli, F. Trotta, U. Dianzani, C. Dianzani, In Vitro and In Vivo Therapeutic Evaluation of Camptothecin-Encapsulated β-Cyclodextrin Nanosponges in Prostate Cancer, J Biomed Nanotechnol 12(1) (2016) 114–27.

5 Inorganic Nanoparticles in Anti-angiogenic Cancer Therapy

The tumor is a broad concept that refers to a group of illnesses characterized by unregulated, spontaneous cell growth and metastatic spread [1]. Cancer is the second most significant risk factor for mortality worldwide and one of the most serious human healthcare issues [2]. By the end of 2021, the American Cancer Society predicts that 1.9 million additional cases will have been diagnosed [3]. Traditional therapeutic strategies are surgical, chemotherapeutic, radiation treatment, targeted therapy, immunotherapy, and hormonal treatment [4, 5]. Although chemotherapeutic and radiotherapy treatments are capable of cytotoxicity and cytostasis [6], they are frequently associated with severe harm ful consequences and a significant probability of reactivation [7]. Bone marrow inhibition, gastric and skin diseases, hair loss, neuropathies, and tiredness are prevalent adverse events [8]. In addition, certain pharmaceuticals have unique adverse consequences, like anthracyclines and bleomycin produce the results respiratory and cardiotoxicity [9].

Over many decades, numerous attempts have been made to identify multiple cancer risk markers [10]. The genesis of several malignancies has been linked to particular environmental (acquisition variables) like radiation and pollutants. An addictive personality, such as poor eating habits, cigarette use, nicotine, anxiety, and insufficient regular exercise, significantly influences tumor risk management [11]. Whereas these extrinsic variables have been identified as critical drivers of malignancy, the role of proto-oncogene alterations, tumor suppressor protein expression profiles, and DNA repairing proteins has been difficult to quantify. Inherited genes are responsible for just 5–10% of cancer cases [10, 12].

Angiogenesis is also a major factor in cancer growth and metastasis [13]. Angiogenesis is the process of forming new blood vessels during growth, propagation, and healing, which pro-angiogenic and anti-proliferative agents regulate. Pathological angiogenesis is critical, particularly in the setting of malignant illness [14]. Cancer cells multiply fast and require a constant delivery of oxygen and nutrients, necessitating a continuous necessity

S. Mukherjee and V. S. Madamsetty, *Nanoparticles in Angiogenesis and Cancer*,
Synthesis Lectures on Biomedical Engineering,
https://doi.org/10.1007/978-3-031-11284-3_5

for invading blood vessels [15]. One of the characteristics of tumors is angiogenesis [16]. Attacking the angiogenic pathway is a rational strategy for treating many cancers, and several anti-angiogenic therapy drugs have been created and evaluated in experimental studies [17]. Dr. Folkman developed a concept of anti-angiogenic treatment after discovering that cancers need constant oxygen and food intake [18]. The neogenesis of cancer vessels aids in the tumor's continual development. Angiogenesis suppression may result in tumor starvation and tumor cell mortality [19, 20]. The fundamental process of new angiogenesis is the development of new blood vessels via pre-existing vasculature [21]. Angiogenic blood vessel creation may take place via three fundamental mechanisms: sprouting angiogenesis, vasculogenesis, and intussusception [22]. Although sprouting angiogenesis is thought to be the most important mechanism of blood vessel development in tumors, data reveals that a non-angiogenic condition known as vascular co-option is also essential [23, 24]. Cancer cells choose vascular co-option at various anatomical places by stealing pre-existing capillaries from the healthy surrounding tissue [23, 25]. Hypoxia is a critical element in tumor angiogenesis [26]. Cancer cells proliferate quickly, are hypoxic, and release vascular endothelial growth factor A (VEGFA) [27]. Pro-angiogenic substances are produced by both tumor cells and tumor-associated stromal cells. The VEGF receptors are expressed by vascular endothelium in the microenvironment [28]. The endothelial cells may detect the VEGF differential through the VEGF–VEGFR association, producing new epithelial sprouts towards the tumor cell [29].

Another major problem in cancer treatment is early detection. Various imaging techniques such as X-ray-based computer-assisted tomography (CT), single-photon emission tomography, positron emission tomography (PET), and magnetic resonance imaging (MRI) are available but still not effective in early detection [30, 31]. Therefore it is urgently needed to identify efficient materials with angiogenesis targets for cancer imaging and therapy.

Nanomaterials or nanoparticles (NPs) are entities with a diameter of lesser than 200 nm with distinctive features not seen in mass substances of the same substance [32, 33]. These could be classed as 0D, 1D, 2D, or 3D nanoparticles based on their general appearance [34]. The surface covering, the outer surface, and the nucleus, which is primarily the center component of the NPs and is typically referred to as the NP itself, constitute the essential structure of nanoparticles [35]. These substances have acquired a lot of relevance in interdisciplinary domains due to their extraordinary qualities, such as more significant surface area: volume fraction, sub-micron dimension, and improved targeting capability. The discovery of deep tissue diffusion of NPs has resulted in an enhanced permeation and retention (EPR) impact [36].

Furthermore, by successfully bridging epithelium fenestration, surface features influence pharmacokinetics and half-life [37]. For instance, NPs covered with the hydrophilic polymer polyethylene glycol (PEG) reduce opsonization and immunologic elimination (Prokop et al. 2008) [38]. Manipulation of nanoparticle polymer properties may also be used to improve medication or bioactive constituent release kinetics [39]. Their unique features regulate the pharmacological potential of NPs in cancer therapeutic interventions and therapy [39].

5.1 Anti-angiogenic Treatments for Cancer

Angiogenesis antagonists work by targeting existent tumor-infiltrating blood capillaries and preventing the creation of the new vascular system, halting tumor metabolism and development [40]. Angiogenesis inhibitors may induce the degradation of immature angiogenic arteries or stop the angiogenic shift, inhibiting capillary metastasis, focusing on the primary mechanism of neo-angiogenesis in initial malignancies [41]. Some prior studies demonstrate that anti-angiogenic therapy significantly improves anticancer effectiveness when combined with chemotherapeutics since these drugs can slow but not stop tumor development [42]. Angiogenesis inhibitors are divided into two categories: direct inhibitors and medicines that target key angiogenesis components & Indirect antagonists, which attack tumor cells or stromal cells and alter angiogenesis controllers, constitute the second type. [43] The anti-angiogenic impact of direct angiogenesis inhibitors is achieved by interacting with angiogenesis stimulants such as VEGF, bFGF, and PDGF [42]. In 2004, the FDA authorized bevacizumab as the first anti-angiogenic therapy for the management of colon cancer [44]. Anti-angiogenic therapy is the 4th type of treatment for cancer, and it works by targeting microvascular vascular endothelium in the tumor microenvironment rather than the tumor itself [45]. Fast-growing tumors require more microvasculature and are thus more vulnerable to anti-angiogenic treatment [46].

5.2 Inorganic Nanoparticles: Approach to Improve Anti-angiogenic Therapy

The majority of NPs fall into two main categories: organic (NPs made up of organic elements) and inorganic (Nanoparticle made up of inorganic components) [37]. The inorganic NPs are made up of two parts: a nucleus/core (which contains metals such as silver, gold, quantum dots, carbon composites, and iron oxide) and a covering/shell (which contains no metals) (organic polymers or metals that defend the core/nucleus from chemical contacts or assists as a substrate for conjugation with biomolecules) [47]. Carbon nanotubes, graphene, fullerene, and graphene oxide are allotropes of carbon, such as diamond, and may not be considered organic but instead considered inorganic materials, so we include these NPS here. Several kinds of inorganic-based nanoparticles have been explored, including a metallic, metal oxide, and magnetic nanoparticles discussed below.

5.3 Metallic Nanoparticles

Due to their outstanding optical, magnetic, and photodynamic capabilities, nanostructured materials are often investigated in biological imaging and targeted drug delivery [48]. Angiogenesis, which is necessary for physiological functions, is also crucial in pathological situations like tumor development and metastasis [49]. Gold NPs, silver NPs, iron-based NPs, and copper NPs are among the most studied metallic NPs developed for anti-angiogenic therapy [50]. Because the size and surface features of gold NPs may be readily regulated, they are employed as intracellular targeted medication vehicles [51]. Furthermore, their visible light attenuation behavior allows tracking NP inside the cells. For examples. The use of Anti-HER2 functionalized gold-on-silica nanoshells to target HER2 positive breast cancer cells has been demonstrated [52].

Further, Combidex®, an iron oxide NP, showed excellent results in vitro and in vivo and is now being tested for nodal metastasis in late-stage clinical trials [53]. Feraheme® is also an iron oxide-based nanoparticle developed by using ferumoxytol and used to cure iron deficiency anemia [54]. The FDA licensed this drug in June 2009 to treat nodal metastases in prostate and testicular malignancy [55, 56].

Plant extracts are a cost-effective and environmentally sound technique to make nanoparticles and are used as anti-angiogenic therapeutics [57]. Baharara et al. 2014, studied the anti-angiogenesis capabilities of silver nanoparticles synthesized on chick chorioallantoic membrane using Saliva officinalis extract [58]. The color shift from yellow to brown confirmed the metal nanoparticle synthesis after 3 h at 37 °C stirrings. These nanoparticles showed better results than other controls regarding the quantity, lengths of the vascular system, the CR (Crown Rump), and mass of the embryos [58]. Further, the angiogenesis suppressive impact of AgNPs was shown by a reduction in hematocrit in the samples treated than controls [58]. Balakrishnan and co-workers demonstrated that naturally available photochemical quercetin conjugated gold nanoparticles showed inhibition of epithelial-mesenchymal transition, tumor angiogenesis, and tumor metastasis via EGFR/VEGFR-2-pathway in in vitro and in vivo breast cancer tumor model (Fig. 5.1) [59].

In another work, Luo et al. 2019, demonstrate thatBgelatin-capped silver nanoparticles (G-Ag NPs) are used to treat S. aureus-induced keratitis. Silver nitrate, maltose, and gelatin were used to develop these G-Ag NPs, and gelatin was encapsulated on Ag NPs (14 nm) in situ [60]. Further, G-Ag NPs outperform uncapped Ag NPs regarding persistence and antimicrobial activities toward S. aureus. G-Ag NPs also impede human umbilical vein endothelial cells' growth, motility, and cellular proliferation. According to the authors, they caused significant angiogenesis disruption in chick chorioallantoic membrane and rabbit ocular neovascularization [60]. Moreover, intrastromal delivery of biodegradable and biocompatible G-Ag NPs reduces microbial keratitis and microbial

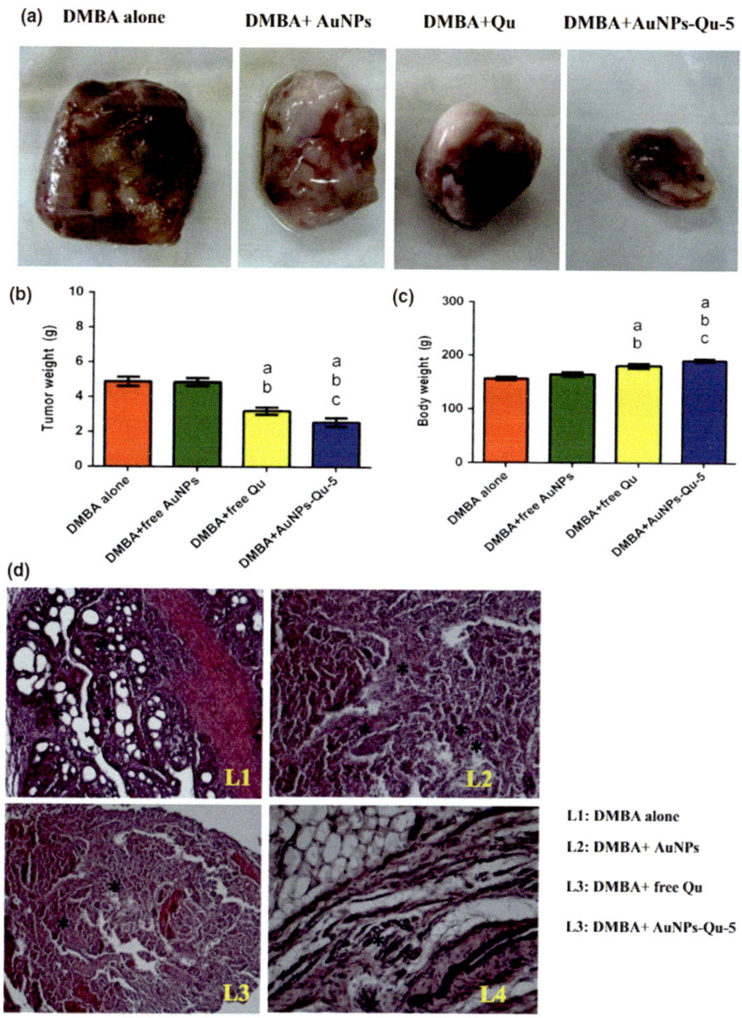

Fig. 5.1 Effect of free Qu and AuNPs-Qu-5 on DMBA- induced mammary carcinoma in Sprague-Dawley rats. AuNPs-Qu-5 inhibited the DMBA- induced tumor growth in Sprague- Dawley rats: **a** Representative photographs of breast tumors in each group; **b** weight of breast tumors in each group; **c** body weight of all the animals in each group. "a" DMBA alone versus others; "b" DMBA induced animals + free AuNPs versus others; and "c" DMBA induced animals + free Qu. **d** Effect of free Qu and AuNPs- Qu-5 on histopathological examination. **d** Histopathological examination of DMBA-induced breast cancer in Sprague- Dawley rats. Histopathological changes in the mammary tissues of the cancer-induced vehicle and experimental animals (hematoxylin and eosin, 10 ×). L1: Cancer-induced breast cancer animals show the extensive solid areas and several neoplastic cells lobular structural disruption; L2: CI + free AuNPs show extensive solid tumor; L3: free quercetin- treated animal shows few amounts of neoplastic structure; L4: AuNPs-Qu-5-treated animal shows normal mammary epithelial cells appearance [59]. *Represent neoplastic cells in DMBA- induced group

infection-induced corneal angiogenesis in rabbit eyeballs caused by S. aureus. According to findings, g-Ag NPs are a potential dual-function (antibacterial and anti-angiogenic) nanotherapeutic for experimental therapy of eye-related microbial pathogens [60].

Hajebi and colleagues (2019) have discovered that bio-synthesized silver nanoparticles (Ag-NPs) can reduce angiogenesis and oxidative stress in human tissue [61]. According to their findings of this research, rapeseed flower pollen (RFP) aqueous extract was used to initiate the production of RFP-silver nanoparticles (RFP/Ag-NPs). Additionally, the antioxidant, anti-angiogenesis, and cytotoxicity of the "RFP/Ag-NPs" against "MDA-MB-231 and MCF7, and carcinoma cell lines and normal human skin fibroblast HDF were evaluated and compared to each other. Their results suggested that the green-synthesized RFP/Ag-NPs, which act as a safe and effective apoptosis inducer and a potent antioxidant agent, can reduce breast cancer carcinogenesis by down-regulating VEGF expression [61]. Nevertheless, more studies are needed to determine the cell selectivity of RFP/Ag-NPs and the pharmacological dosages that are effective in vivo circumstances [61]Mukherjee et al. demonstrated the dose-dependent manipulation of anti-angiogenesis and angiogenesis utilizing the graphene oxides treatment [62]. These NPs worked via modulating ROS in in vitro scratching assay, endothelial cell proliferation assay, tube formation assay, and chicken embryonic angiogenesis assay [62].

Silver and gold nanoparticles have distinct biological and physical features that have been intensively investigated for their potential use in biological and medicinal purposes [63, 64]. For the most part, gold and silver nanoparticles are produced by chemical reducing agents that use surplus hazardous molecules that must be eliminated before they can be used for biological reasons [65]. To synthesize metal nanoparticles to investigate their possible impacts on angiogenesis regulation, Kemp et al. 2009, used an environmentally friendly technique requiring just a solitary synthesis process [66]. These nanostructures were created by converting silver nitrate and gold chloride with "diaminopyridinyl (DAP)-derivatized heparin (HP) polysaccharides", which were then dried and powdered. The gold and silver nanoparticles with DAPHP decreased the suppression of fundamental fibroblast growth factor (FGF-2)-induced angiogenesis. DAPHP-reduced metal nanoparticles may promise to treat pathogenic angiogenesis accelerated conditions such as cancer and inflammatory disorders [66].

Angiogenesis-inhibiting properties of AuNPs were established in a number of research investigations [67]. The survival, migratory, and colony formation of human umbilical vein endothelial cells (HUVECs) were all inhibited by AuNPs. The anti-angiogenic effects of AuNPs on human umbilical vein endothelial cells (HUVECs) were shown to be dose-dependent [68, 69]. In addition, AuNPs have been demonstrated to prevent the migratory of human retinal microvascular endothelial cells (HRMECs) triggered by VEGF-A in vitro [70]. Additional evidence of AuNPs' anti-angiogenic effect has been obtained via ex vivo studies like the chick chorioallantoic membrane model (CAM) [71]. AuNPs hindered the creation of a new vascular system in the CAM system, and they also lowered the total tubule structure, the overall vessel size, and the number of vascular connections. AuNPs have also been an anti-angiogenic effect when administered in vivo in various animal models [72].

AuNPs lowered vascular volume and transparency in diverse tumor models while maintaining vessel strength. The administration of AuNPs restored normal tumor vasculature in nude mice infected with SW620 human colon carcinoma [73]. Using the CD31 vascular marker to stain tumor tissues, it was discovered that AuNPs decreased vascular volume. Aside from this, AuNPs enhanced vascular integrity by boosting pericyte coverage, as assessed by the identification of -smooth muscle actin in the tumor vasculature after 6–9 days of therapy after administration [73]. The anti-angiogenic effect of AuNPs was proven in an animal model of cancer, with decreased vascular volume, decreased permeability, and better vasculature shape and circulation being the primary outcomes. When comparing AuNPs-treated mice to vehicle-treated controls, normalization of tumor vasculature was related to a reduction in lung metastasis [74]. As shown by staining for the marker CD34, the anti-angiogenic effect of AuNPs in an animal study of liver cancer resulted in an impressive reduction in vasculature volume in animals given the treatment [74]. AuNPs also impacted capillary permeability in a mouse ovarian tumor model, as shown by less fluid accumulated in the peritoneal cavity of the mice treated with nanoparticles [74]. Mukherjee et al. showed the intrinsic property of gold nanoparticles [75]. They showed that gold nanoparticles explicitly bound vascular permeability factor/vascular endothelial growth factor-165 and basic fibroblast growth factor, two endothelial cell mitogens and angiogenesis mediators, resulting in endothelial/fibroblast cell proliferation inhibition in vitro VEGF-induced permeability as well as angiogenesis in vivo (Fig. 5.2) [75]. Additionally, a decrease in angiogenesis was found in a nude mouse ear model injected with gold nanoparticles [75]. In summary, metal nanoparticles showed vast applications in anti-angiogenesis therapy.

5.4 Metal Oxide Nanoparticles

Expanding nanomaterials with high biomedical applications with low toxicity is essential to obtain more efficient imaging and therapy tools [76]. The metal oxide nanoparticles are among the most comprehensively manufactured nanomaterials because of their unique imaging and therapeutic applications properties [77, 78]. Metal targets such as iron (Fe), zinc (Zn), aluminum (Al), titanium (Ti), zinc (Zn), nickel (Ni), copper (Cu), gadolinium (Gd), tin (Sn), cobalt (Co), cerium (Ce), magnesium (Mg), bismuth (Bi), yttrium (Y), etc. have been used to make a variety of metal oxide nanoparticles and are described below.

Mani et al. (2018) produced gelatin-coated zinc oxide NPs through the co-precipitation method [79]. They found that at 50 g/ml, Ge-ZnO NPs efficiently prevented the biofilm

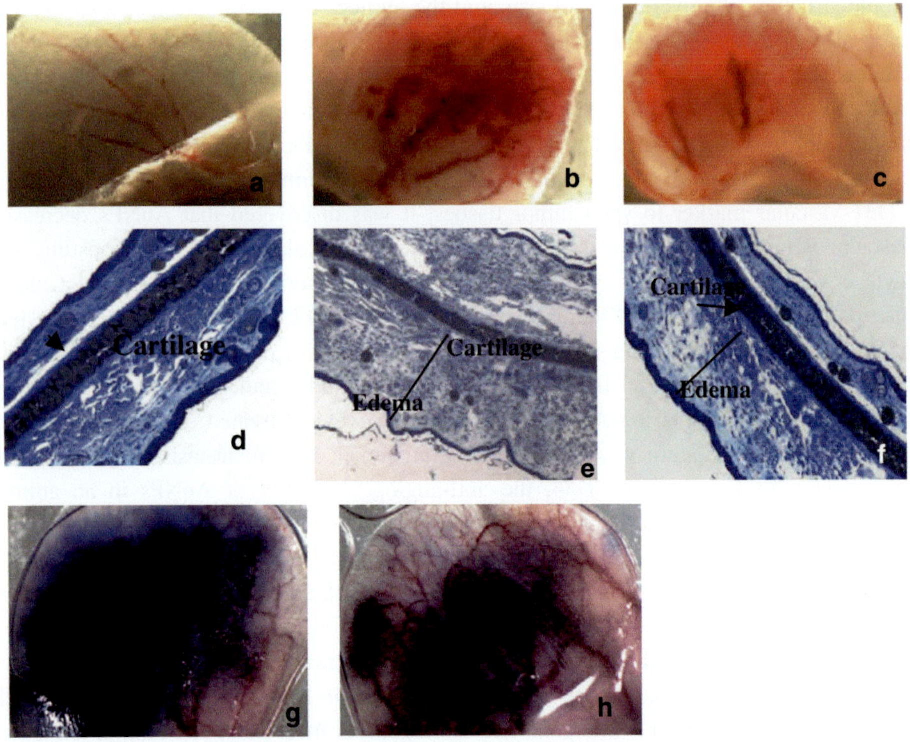

Fig. 5.2 Effect of nanogold on angiogenesis in vivo in the ears of nude mice. Gross appearance of angiogenesis seven days after injection of nanogold only (**a**), Ad-VEGF only (**b**), nanogold and Ad-VEGF (**c**). Giemsa stained 1 Amol/L epon sections of the ears were photographed at 10X magnifications, nanogold only (**d**), Ad-VEGF only (**e**), nanogold and Ad-VEGF (**f**). Effect of nanogold on permeability, Ad-VEGF only (**g**), nanogold, and Ad-VEGF (**h**)

formation of the fungus Candida albicans. At 100 g/ml, Ge-ZnO NPs inhibited the hepatocarcinoma cancer cell line proliferation. Furthermore, significant anti-angiogenesis activities were detected in chick embryos for Ge-ZnO NPs and zinc acetate at 50 g/ml compared to gelatin [79].

The chemical explanation behind the vascular toxicity of copper oxide nanoparticles (CuONPs) is unknown. Recently, He and his colleagues successfully proved that the discharge of copper ions and the acquisition of superoxide radicals attributed to CuONPs-induced apoptosis in human umbilical vein endothelial cells (HUVECs) so that CuONPs stimulated oxidative DNA injury and induced apoptosis in HUVECs via copper ions associated p38 MAPK stimulation [80].

In another study, Raj Preeth et al. 2019, used sinapic acid to synthesize copper oxide nanoparticles [81]. This work employs sinapic acid, a natural product, to make persistent copper oxide nanoparticles. Indepth characterization of these NPs was evaluated using

UV–Visible, IR spectroscopy, Transmission Electron Microscopy (TEM), and X-ray photoelectron spectroscopy. These nanomaterials have shown possible cytotoxic effects on MCF7 and MDA-MB231 breast cancer cells. Furthermore, these NPs showed an anti-angiogenic impact on endothelial cells (EA.hy926), indicating that they can minimize angiogenesis in tumors [81].

Another group of scientists investigated the anti-angiogenic properties of ZnWO4 nanoparticles (NPs) [82]. Microwave-assisted hydrothermal synthesis (MAHS) was used to make these nanomaterials, which were then architecturally studied using X-ray diffraction (XRD) and micro-Raman (MR) spectroscopy at 120 °C for 60 min. The findings revealed that nanocrystal ZnWO4 NPs produced by MAHS at low reaction temperatures have good anti-angiogenic effects even at small concentrations. The cytotoxicity of these ZnWO4 NPs was also assessed in vitro [82].

Angiogenesis is a pathophysiologic activity that combines the formation of new bloodstreams from existing ones [83]. Despite indications that cuprous oxide nanoparticles (CO-NPs) may have biological properties, the processes of CO-NPs in angiogenesis have yet to be studied [84]. The researchers examined the anti-angiogenic characteristics of CO-NPs on foremost human umbilical vein endothelial cells (HUVECs). They discovered that CO-NPs could stimulate cell morphology variations and inhibit cell growth, migratory. Colony formation in vitro and in vivo dose-dependent manner and demonstrate for the first time that CO-NPs can function like an anti-angiogenic entity by restricting VEGFR2 expression [84].

Nanoceria (non-cytotoxic cerium oxide nanoparticles) have been shown to control intracellular reactive oxygen species (ROS). Lord and colleagues used the organosilane linker 3-aminopropyltriethoxysilane to functionalize nanoceria (dxRD = 12 nm) with 130 or 880 units of unfractionated heparin [85]. Nanoceria with a modest amount of heparin functionalization scavenged internal ROS to the same amount as unfunctionalized nanoceria, and far more than cells were exposed to medium alone [85]. Heparin-nanoceria were found in both the cytoplasmic and endosomal, while nanoceria was mostly found in the cytoplasm. These findings showed that the amount of heparin surface functionalization on nanoceria nanoparticles impacted their intracellular localization and ROS scavenging capabilities [85]. Furthermore, heparin-nanoceria successfully lowered cellular proliferation, suggesting that they may be used in the future to inhibit angiogenesis in tumors [85]. Similarly, there are several metal oxide nanoparticles developed for anti-angiogenic therapy [86–89].

5.5 Magnetic Nanoparticles

Metal or metal oxide nanoparticles are employed in medication administration, and magnetic NPs are often used in MRI imaging for cancer detection [90, 91]. To improve stability and biocompatibility, they are generally coated with organic compounds such as polymers and fatty acids [92]. Angiogenesis molecular targets are used to detect cancer by using these nanoparticles [93]. For example, Breast cancer may be targeted and imaged using LHRH-conjugated superparamagnetic iron oxide NPs [94]. Furthermore, magnetic NPs are employed in magnetic hyperthermia for cancer cell thermal ablation [95]. Feridex® and Resovist® for liver metastases and colon cancer are two magnetic NPs that are now on the market or in clinical trials [96].

Artemisinin, a natural anti-malarial drug, has anti-proliferative and anti-angiogenic properties in cancer cells while causing very little harm in normal healthy cells [97, 98]. Drug-loaded magnetic nanoparticles might selectively concentrate the drug at the target location using an external magnetic field, reducing the dosages necessary to obtain effective concentrations that could otherwise cause significant adverse effects in normal tissues [97]. Natesan et al. 2017 used the ionic-gelation approach to manufacture artemisinin magnetic nanoparticles utilizing chitosan in this investigation effectively [97]. The artemisinin magnetic nanoparticles created were smoother and circular, with a 349–445 nm size range. The polydispersity index (PDI) and zeta potential of these NPs were 0.373–0.908 and −9.34 to −33.3 mV. The drug entrapment effectiveness of these NPs was 55–62.5%, with 20–25% drug loading capacity. Over 48 h, approximately 62–78% of artemisinin was liberated from the artemisinin magnetic NPs. FITC coupled artemisinin magnetic nanoparticles accumulated in the 4T1 breast cancerous tissue of BALB/c mice when exposed to a medically tolerable externally applied magnetic field [97].

In another study, Abakumov et al. 2015, used thermal decomposition to create ferric oxide (Fe3O4) nuclei covered with bovine serum albumin (BSA) to make nanoparticles with a diameter of 539 nm [99]. To increase colloidal integrity, BSA was extensively cross-linked. The mAbVEGF-targeted BSA-coated magnetic nanoparticles were demonstrated as potential MRI contrast enhancers for glioma imaging after monoclonal antibodies targeting vascular endothelial growth factor (mAbVEGF) were covalently conjugated to BSA using a polyethyleneglycol (PEG) linker [99]. Karimi and his research team created carrageenan-crosslinked magnetic chitosan with various molecular weights as pH-responsive vehicles for the antitumor medication sunitinib [100]. The diameter of magnetic particles is influenced by the molecular mass of chitosan, according to the characterization of magnetic transporters [100]. The diameter of metal particles affects drug entrapment efficiency and discharge efficiency. Correspondingly, Sunitinib encapsulation rates were 62.38, 69.57, and 78.42% for low, medium, and molecular-weight magnetic chitosan vehicles [100]. The release of sunitinib from magnetic chitosan/-carrageenan transporters in vitro was pH-dependent and followed a Fickian process. Sunitinib was effectively emitted into the environment from magnetic transporters at acidic pHs, and the rate of release was diameter and molecular mass-dependent [100] The current magnetic transporters are an excellent choice for antitumor medications with fewer adverse reactions because of the pH-dependent liberation of sunitinib, which has a minimum discharge rate at pH = 7.4 [100].

5.6 Calcium Phosphate Nanoparticles

Calcium phosphate NPs are biocompatible, biodegradable, pH-responsive, do not produce any severe side effects, and are considered ideal drug carriers for delivering anticancer drugs, including anti-angiogenesis agents [101, 102]. These NPs are also employed to transport insulin, human growth hormone, medicines, and contraceptives [103]. They're also used for oligonucleotide and gene delivery [104]. Calcium phosphate NPs have been successfully employed as strategic in intracellular gene delivery when paired with a viral or non-viral carrier [105, 106]. The cytotoxicity of a liposomal nanolipoplex formulation of calcium and glycerol was reduced, and the transfection properties were improved [107]. A recent study by Huang et al. showed that lipid/calcium-phosphate nanoparticles were efficient in delivering VEGF (vascular endothelial growth factor) siRNA in orthotopic hepatocellular carcinoma (HCC) tumors in the murine liver to inhibit tumor growth [106]. These NPs were conjugated with galactoside derivatives to target asialoglycoprotein receptors (ASGPR) on the HCC cells and delivered siRNA selectively [106]. Their findings in both in vitro and in vivo showed that the β-D-galactoside conjugated NPs better HCC targeting efficiency than non-conjugated NPs and released VEGF siRNA to the cytosol [106]. siRNA downregulated VEGF expression in tumor cells and prevented tumor grwth by an anti-angiogenesis mechanism [106].

5.7 Silica Nanoparticles

Silica, which is found in many natural compounds, has only lately been investigated in terms of life. By surface functionalization of the NPs membrane with amino-silicones, silica NPs are extensively employed to transport genes (Katragadda et al. 2021). N-(6–aminohexyl)–3–aminopropyl–trimethoxysilane functionalized silica NPs have shown high efficacy in the transfection of Cos-1 cells while posing low toxicity and are now available commercially (Kneuer et al. 2000). Because of their superior pharmacological features, mesoporous silica nanoparticles are regarded as one of the most acceptable drug carriers. Recent study provides a novel strategy of encapsulating bevacizumab to protect and deliver it, increasing the time between administration and formulation shelf-life. It showed a promising drug delivery alternative to anti-angiogenic therapy.

5.8 Carbon-Based Nanoparticles

Graphene, carbon nanotubes, fullerenes, carbon nanohorns, and graphene oxide are several types of carbon NPs considered inorganic nanoparticles as these are not authentic organic materials [108]. Due to their optical, mechanical, and electrical qualities and their biocompatibility, they have been extensively used in medical fields [109]. Carbon Nanoparticles may encapsulate medicines by π-π stacking due to their intrinsic hydrophobic character [110]. Despite the fact that they are all carbon-based, they differ in structure, morphology, and characteristics [111]. For example, Graphene is a two-dimensional crystal made out of an sp2-hybridized carbon sheet with exceptional mechanical, electrochemical, and drug-loading capabilities [112]. Graphene may also be split into three categories depending on its composition, characteristics, and composition: Single-layer graphene, graphene oxide (GO), reduced graphene oxide (rGO), and multi-layer graphene are the four types of graphene [113]. Because of its potential to address hypoxia [114] and irregular angiogenesis in TMEs, GO, and rGOs are frequently employed [115]. In preclinical studies of breast carcinoma, research has revealed that GO-doxorubicin has more potent antitumor activity [116].

Fullerenes are enormous carbon-cage structures made up of carbon allotropes in various shapes, including spheres, ellipsoids, and tubes [117]. Because they exhibit specific structural, structural, biochemical, and electrochemical properties, they are the most researched nanocarriers [118]. Due to prolonged conjugation and the capacity to absorb light, they are employed in photodynamic treatment since they have a triple yield and create oxygen species [119]. On tumor cells, PEG-modified fullerenes demonstrated potential photodynamic effects [120]. Carbon nanotubes (CNTs) are cylindrical tubes most often referred to as graphene rolls [121]. They were first found in the late 1980s [121]. They are divided into two categories: There are two types of carbon nanotubes: single-walled and multi-walled. Because they are carbon-based, they may elicit an immunological reaction by associating with immune cells, preventing tumor development [121]. They've been employed as DNA transport vehicles and for thermal ablation treatment in the past. A fluorescent single-walled CNT enclosing doxorubicin is used to treat colon cancerous cells. These CNTs create a compound that is successfully absorbed by cancer cells, causing doxorubicin to be released intracellularly while the CNTs are kept in the cytoplasm [122].

Das et al. 2020, examined the use of a mixture of "cocktail chemotherapy" (CCT), "photothermal treatment" (PTT), and angiogenesis suppression to fight the key issues of multidrug tolerance and non-targeted therapeutic agents that are common in traditional treatment of cancer [123]. Carbon nanotubes (CNTs) were functionalized with a strategically altered carbohydrate (-Cyclodextrin-CD) generated pH and temperature-sensitive polymer to create an injectable nanocarrier [123].The results showed a synergistic impact in initiating anticancer effects in vitro and ex-ovo reduced expression of angiogenesis-related growth factors. The nanoparticle made with the reasoning for MDR reversal may lead to accelerated cancer cell death through several methods, according to in-vivo investigations [123].

Angiogenesis is necessary for appropriate wound repair and development [83]. An inequality of the growth factors associated with this mechanism, on the other hand, accelerates the progression of a variety of disorders, notably cancer, ophthalmology, and inflammatory diseases. Restricting angiogenesis by interacting with its mechanism is a potential strategy for slowing disease development [124]. Murugesen and coworkers (2007) investigated the anti-angiogenic impacts of several carbon materials, including graphite, multiwalled carbon nanotubes, and fullerenes, in VEGF (vascular endothelial growth factor) and fibroblast growth factor(FGF2)-induced angiogenesis in the "chick chorioallantoic membrane" (CAM) model [125]. All of the carbon materials examined demonstrated significant anti-angiogenic action towards FGF2- or VEGF-induced angiogenesis in the CAM model [125]. In the lack of substantial development agents, those carbon compounds had no meaningful influence on basal angiogenesis [125].

5.9 Quantum Dots

Quantum dots, generally nanoscale semiconductors with broad absorption spectra, limited emission regions, and extraordinary photostability, are commonly utilized in bioimaging [126]. Carbon quantum dots are classified as (1) graphene quantum dots, (2) nanodiamond quantum dots, and (3) carbon quantum dots [127]. Aside from bioimaging, quantum dots are now being studied to treat cancer [128]. Graphene quantum dots are the most often employed due to their intrinsic biocompatibility and quick elimination. Quantum dots aptamer-doxorubicin conjugation targets prostate cancer cells [128]. The biggest obstacle is the lack of an optimal procedure for generating quantum dots.

Some nanoparticles have been demonstrated to have antitumor capabilities. However, one of the obstacles is a lack of specificity, let alone targeted reduction of tumor progression by modulating the cellular microenvironment. Chen and colleagues (2021) reported for the first time that "carbon quantum dots/Cu2O composite (CQDs/Cu2O)" effectively suppressed ovarian cancer SKOV3 cells by attacking cellular microenvironment factors such as matrix metalloproteinases, angiogenic cytokines, and the cytoskeleton [129]. CQDs/Cu2O were shown to have anticancer characteristics towards SKOV3 cells, with an IC50 of 0.85 g mL-1, nearly twofold less than other tumor cells examined and nearly 12-fold lower than normal cells. As a result, CQDs/Cu2O preferentially induced ovarian cancer SKOV3 cellular damage primarily by lowering the levels of MMP-2, MMP-9, F-actin, and VEGFR2, while CQDs/Cu2O induced SKOV3 apoptosis through S phase cell cycle arrest. These results suggest a novel use for CQDs/Cu2O composites as prospective pharmacological treatments in SKOV3 ovarian cancer cells [129].

This study reviewed the latest breakthroughs in using metallic nanoparticles, metal oxide nanoparticles, magnetic nanopartciles and some carbon-based analogs nanoparticles as anti-angiogenesis nanomaterials for cancer treatment. The necessity of pharmacological investigations of nanoparticles as therapies was also recommended to attain maximal therapeutic effectiveness at the nano/bio interaction. Furthermore, scientists need a thorough understanding of nanomaterial's cyto- and genotoxicity and their better construction. The majority of the metallic or metal-containing nanoparticles studied were found to have minimal or no toxic effect. Nevertheless, instead of scientists/toxicologists, nanotechnologists conducted most of such research. Additionally, due to the complicated nature of biological systems, it is difficult to anticipate the behavior and destiny of nanoparticles in vivo. It is necessary to conduct a systematic study of whole-body impacts, taking into account exposure dose, deposition and elimination, tissue/organ distributions, and possible chronic effects. Effective removal of off-target/unbound nanomaterials for healthcare quality and therapeutic implementation is crucial. The bioavailability and cytotoxicity of nanomaterials must be investigated in the targeted tissue and other potential aggregation locations such as the liver and spleen. Scale-up approaches must also be assessed earlier in the strategy. Most of the work described has been done at the laboratory bench to evaluate proof-of-principle notions. Nevertheless, scaling up nanoparticle production under the same manufacturing circumstances might be a significant stumbling block to their effective translation. The route to a smooth adaptation to clinical application gets considerably easy when nanomaterials are correctly developed and produced with thoughtful consideration of these issues.

References

1. E.J. Ge, A.I. Bush, A. Casini, P.A. Cobine, J.R. Cross, G.M. DeNicola, Q.P. Dou, K.J. Franz, V.M. Gohil, S. Gupta, Connecting copper and cancer: from transition metal signalling to metalloplasia, Nature Reviews Cancer (2021) 1–12.
2. C.J. Murray, A.Y. Aravkin, P. Zheng, C. Abbafati, K.M. Abbas, M. Abbasi-Kangevari, F. Abd-Allah, A. Abdelalim, M. Abdollahi, I. Abdollahpour, Global burden of 87 risk factors in 204 countries and territories, 1990–2019: a systematic analysis for the Global Burden of Disease Study 2019, The Lancet 396(10258) (2020) 1223–1249.
3. R.L. Siegel, K.D. Miller, H.E. Fuchs, A. Jemal, Cancer Statistics, 2021, CA: A Cancer Journal for Clinicians 71(1) (2021) 7–33.
4. Z. Abbas, S. Rehman, An overview of cancer treatment modalities, Neoplasm 1 (2018) 139–157.
5. H. Yildizhan, N.P. Barkan, S.K. Turan, Ö. Demiralp, F.D.Ö. Demiralp, B. Uslu, S.A. Ōzkan, Treatment strategies in cancer from past to present, Drug Targeting and Stimuli Sensitive Drug Delivery Systems, Elsevier 2018, pp. 1–37.
6. L. Zitvogel, L. Apetoh, F. Ghiringhelli, G. Kroemer, Immunological aspects of cancer chemotherapy, Nature reviews immunology 8(1) (2008) 59–73.
7. R.G. Armando, D.L. Mengual Gómez, D.E. Gomez, New drugs are not enough-drug repositioning in oncology: An update, International Journal of Oncology 56(3) (2020) 651–684.
8. P.S. Sisodiya, Plant derived anticancer agents: a review, Int. J. Res. Dev. Pharm. Life Sci 2(2) (2013) 293–308.
9. A.A. Alghorabi, A.M. Kabel, M. Elmaaboud, Doxorubicin: Insights into Dynamics, Clinical Uses and Adverse Effects, J. Cancer Res 7 (2019) 17–20.
10. S. Wu, W. Zhu, P. Thompson, Y.A. Hannun, Evaluating intrinsic and non-intrinsic cancer risk factors, Nature communications 9(1) (2018) 1–12.
11. S. Gavas, S. Quazi, T.M. Karpiński, Nanoparticles for cancer therapy: current progress and challenges, Nanoscale Research Letters 16(1) (2021) 1–21.

12. .H. Moxley, D. Reisman, Context is key: Understanding the regulation, functional control, and activities of the p53 tumour suppressor, Cell biochemistry and function 39(2) (2021) 235–247.
13. J. Folkman, Role of angiogenesis in tumor growth and metastasis, Seminars in oncology, Elsevier, 2002, pp. 15–18.
14. M.R. Javan, A. Khosrojerdi, S.M. Moazzeni, New insights into implementation of mesenchymal stem cells in cancer therapy: prospects for anti-angiogenesis treatment, Frontiers in oncology (2019) 840.
15. S. Jafari Nivlouei, M. Soltani, E. Shirani, M.R. Salimpour, R. Travasso, J. Carvalho, A multiscale cell-based model of tumor growth for chemotherapy assessment and tumor-targeted therapy through a 3D computational approach, Cell proliferation 55(3) (2022) e13187.
16. D. Gems, J.P. de Magalhães, The hoverfly and the wasp: A critique of the hallmarks of aging as a paradigm, Ageing research reviews 70 (2021) 101407.
17. F. Lopes-Coelho, F. Martins, S.A. Pereira, J. Serpa, Anti-angiogenic therapy: Current challenges and future perspectives, International Journal of Molecular Sciences 22(7) (2021) 3765.
18. J. Folkman, P. Hahnfeldt, L. Hlatky, Cancer: looking outside the genome, Nature Reviews Molecular Cell Biology 1(1) (2000) 76–79.
19. A.-P. Nikolakopoulos, The history and future perspectives of anti-angiogenics, (2021).
20. J. Jászai, M.H. Schmidt, Trends and challenges in tumor anti-angiogenic therapies, Cells 8(9) (2019) 1102.
21. M. De Palma, D. Biziato, T.V. Petrova, Microenvironmental regulation of tumour angiogenesis, Nature Reviews Cancer 17(8) (2017) 457–474.
22. P. Carmeliet, R.K. Jain, Molecular mechanisms and clinical applications of angiogenesis, Nature 473(7347) (2011) 298–307.
23. E.A. Kuczynski, P.B. Vermeulen, F. Pezzella, R.S. Kerbel, A.R. Reynolds, Vessel co-option in cancer, Nature reviews Clinical oncology 16(8) (2019) 469–493.
24. D. Ribatti, F. Pezzella, Overview on the different patterns of tumor vascularization, Cells 10(3) (2021) 639.
25. S.I. Stiver, Angiogenesis and its role in the behavior of astrocytic brain tumors, Front Biosci 9(1-3) (2004) 3105–23.
26. A.R. Farina, L. Cappabianca, M. Sebastiano, V. Zelli, S. Guadagni, A.R. Mackay, Hypoxia-induced alternative splicing: the 11th Hallmark of Cancer, Journal of Experimental & Clinical Cancer Research 39(1) (2020) 1–30.
27. M. Geindreau, F. Ghiringhelli, M. Bruchard, Vascular endothelial growth factor, a key modulator of the anti-tumor immune response, International Journal of Molecular Sciences 22(9) (2021) 4871.
28. P.M. Lacal, G. Graziani, Therapeutic implication of vascular endothelial growth factor receptor-1 (VEGFR-1) targeting in cancer cells and tumor microenvironment by competitive and non-competitive inhibitors, Pharmacological research 136 (2018) 97–107.
29. L. Jakobsson, C.A. Franco, K. Bentley, R.T. Collins, B. Ponsioen, I.M. Aspalter, I. Rosewell, M. Busse, G. Thurston, A. Medvinsky, Endothelial cells dynamically compete for the tip cell position during angiogenic sprouting, Nature cell biology 12(10) (2010) 943–953.
30. X. Wang, L. Yang, Z. Chen, D.M. Shin, Application of nanotechnology in cancer therapy and imaging, CA: a cancer journal for clinicians 58(2) (2008) 97–110.
31. M.J. Willemink, A. Varga-Szemes, U.J. Schoepf, M. Codari, K. Nieman, D. Fleischmann, D. Mastrodicasa, Emerging methods for the characterization of ischemic heart disease: ultrafast Doppler angiography, micro-CT, photon-counting CT, novel MRI and PET techniques, and artificial intelligence, European Radiology Experimental 5(1) (2021) 1–14.

32. P.N. Sudha, K. Sangeetha, K. Vijayalakshmi, A. Barhoum, Nanomaterials history, classification, unique properties, production and market, Emerging applications of nanoparticles and architecture nanostructures, Elsevier2018, pp. 341–384.
33. L.H. Madkour, Introduction to nanotechnology (NT) and nanomaterials (NMs), Nanoelectronic Materials, Springer2019, pp. 1–47.
34. A.B. Asha, R. Narain, Nanomaterials properties, Polymer Science and Nanotechnology, Elsevier2020, pp. 343–359.
35. R. Rampado, S. Crotti, P. Caliceti, S. Pucciarelli, M. Agostini, Recent advances in understanding the protein corona of nanoparticles and in the formulation of "Stealthy" Nanomaterials, Frontiers in Bioengineering and Biotechnology 8 (2020) 166.
36. M.A. Subhan, S.S.K. Yalamarty, N. Filipczak, F. Parveen, V.P. Torchilin, Recent Advances in Tumor Targeting via EPR Effect for Cancer Treatment, J Pers Med 11(6) (2021) 571.
37. I. Khan, K. Saeed, I. Khan, Nanoparticles: Properties, applications and toxicities, Arabian Journal of Chemistry 12(7) (2019) 908–931.
38. J.S. Suk, Q. Xu, N. Kim, J. Hanes, L.M. Ensign, PEGylation as a strategy for improving nanoparticle-based drug and gene delivery, Adv Drug Deliv Rev 99(Pt A) (2016) 28–51.
39. J.K. Patra, G. Das, L.F. Fraceto, E.V.R. Campos, M.D.P. Rodriguez-Torres, L.S. Acosta-Torres, L.A. Diaz-Torres, R. Grillo, M.K. Swamy, S. Sharma, S. Habtemariam, H.-S. Shin, Nano based drug delivery systems: recent developments and future prospects, J Nanobiotechnology 16(1) (2018) 71–71.
40. H. Saman, S.S. Raza, S. Uddin, K. Rasul, Inducing Angiogenesis, a Key Step in Cancer Vascularization, and Treatment Approaches, Cancers (Basel) 12(5) (2020) 1172.
41. D.R. Bielenberg, B.R. Zetter, The Contribution of Angiogenesis to the Process of Metastasis, Cancer J 21(4) (2015) 267–273.
42. J. Ma, D.J. Waxman, Combination of antiangiogenesis with chemotherapy for more effective cancer treatment, Mol Cancer Ther 7(12) (2008) 3670–3684.
43. A.M. Al-Abd, A.J. Alamoudi, A.B. Abdel-Naim, T.A. Neamatallah, O.M. Ashour, Anti-angiogenic agents for the treatment of solid tumors: Potential pathways, therapy and current strategies - A review, J Adv Res 8(6) (2017) 591–605.
44. B. Al-Husein, M. Abdalla, M. Trepte, D.L. Deremer, P.R. Somanath, Antiangiogenic therapy for cancer: an update, Pharmacotherapy 32(12) (2012) 1095–1111.
45. D. Ribatti, T. Annese, S. Ruggieri, R. Tamma, E. Crivellato, Limitations of Anti-Angiogenic Treatment of Tumors, Transl Oncol 12(7) (2019) 981–986.
46. D. Fukumura, R.K. Jain, Tumor microvasculature and microenvironment: targets for anti-angiogenesis and normalization, Microvasc Res 74(2-3) (2007) 72–84.
47. J. Jeevanandam, A. Barhoum, Y.S. Chan, A. Dufresne, M.K. Danquah, Review on nanoparticles and nanostructured materials: history, sources, toxicity and regulations, Beilstein J Nanotechnol 9 (2018) 1050–1074.
48. Y. Yang, L. Wang, B. Wan, Y. Gu, X. Li, Optically Active Nanomaterials for Bioimaging and Targeted Therapy, Frontiers in Bioengineering and Biotechnology 7 (2019).
49. N. Nishida, H. Yano, T. Nishida, T. Kamura, M. Kojiro, Angiogenesis in cancer, Vasc Health Risk Manag 2(3) (2006) 213–219.
50. A. Andleeb, A. Andleeb, S. Asghar, G. Zaman, M. Tariq, A. Mehmood, M. Nadeem, C. Hano, J.M. Lorenzo, B.H. Abbasi, A Systematic Review of Biosynthesized Metallic Nanoparticles as a Promising Anti-Cancer-Strategy, Cancers (Basel) 13(11) (2021) 2818.

51. Q. Gao, J. Zhang, J. Gao, Z. Zhang, H. Zhu, D. Wang, Gold Nanoparticles in Cancer Theranostics, Frontiers in Bioengineering and Biotechnology 9 (2021).
52. Q. Dong, H. Yang, C. Wan, D. Zheng, Z. Zhou, S. Xie, L. Xu, J. Du, F. Li, Her2-Functionalized Gold-Nanoshelled Magnetic Hybrid Nanoparticles: a Theranostic Agent for Dual-Modal Imaging and Photothermal Therapy of Breast Cancer, Nanoscale research letters 14(1) (2019) 235–235.
53. S.M. Dadfar, K. Roemhild, N.I. Drude, S. von Stillfried, R. Knüchel, F. Kiessling, T. Lammers, Iron oxide nanoparticles: Diagnostic, therapeutic and theranostic applications, Adv Drug Deliv Rev 138 (2019) 302–325.
54. Y. Huang, J.C. Hsu, H. Koo, D.P. Cormode, Repurposing ferumoxytol: Diagnostic and therapeutic applications of an FDA-approved nanoparticle, Theranostics 12(2) (2022) 796–816.
55. S. Gavas, S. Quazi, T.M. Karpiński, Nanoparticles for Cancer Therapy: Current Progress and Challenges, Nanoscale research letters 16(1) (2021) 173–173.
56. D.J. Bharali, I.A. Siddiqui, V.M. Adhami, J.C. Chamcheu, A.M. Aldahmash, H. Mukhtar, S.A. Mousa, Nanoparticle delivery of natural products in the prevention and treatment of cancers: current status and future prospects, Cancers (Basel) 3(4) (2011) 4024–4045.
57. P. Kuppusamy, M.M. Yusoff, G.P. Maniam, N. Govindan, Biosynthesis of metallic nanoparticles using plant derivatives and their new avenues in pharmacological applications – An updated report, Saudi Pharmaceutical Journal 24(4) (2016) 473–484.
58. J. Baharara, F. Namvar, T. Ramezani, N. Hosseini, R. Mohamad, Green synthesis of silver nanoparticles using Achillea biebersteinii flower extract and its anti-angiogenic properties in the rat aortic ring model, Molecules 19(4) (2014) 4624–4634.
59. S. Balakrishnan, F.A. Bhat, P. Raja Singh, S. Mukherjee, P. Elumalai, S. Das, C.R. Patra, J. Arunakaran, Gold nanoparticle-conjugated quercetin inhibits epithelial-mesenchymal transition, angiogenesis and invasiveness via EGFR/VEGFR-2-mediated pathway in breast cancer, Cell Prolif 49(6) (2016) 678–697.
60. L.J. Luo, T.Y. Lin, C.H. Yao, P.Y. Kuo, M. Matsusaki, S.G. Harroun, C.C. Huang, J.Y. Lai, Dual-functional gelatin-capped silver nanoparticles for antibacterial and antiangiogenic treatment of bacterial keratitis, Journal of colloid and interface science 536 (2019) 112–126.
61. S. Hajebi, M.H. Tabrizi, M.N. Moghaddam, F. Shahraki, S. Yadamani, Rapeseed flower pollen bio-green synthesized silver nanoparticles: a promising antioxidant, anticancer and antiangiogenic compound, Journal of biological inorganic chemistry : JBIC : a publication of the Society of Biological Inorganic Chemistry 24(3) (2019) 395–404.
62. S. Mukherjee, P. Sriram, A.K. Barui, S.K. Nethi, V. Veeriah, S. Chatterjee, K.I. Suresh, C.R. Patra, Graphene Oxides Show Angiogenic Properties, Advanced Healthcare Materials 4(11) (2015) 1722–1732.
63. S.B. Yaqoob, R. Adnan, R.M. Rameez Khan, M. Rashid, Gold, Silver, and Palladium Nanoparticles: A Chemical Tool for Biomedical Applications, Frontiers in Chemistry 8 (2020).
64. L.A. Dykman, N.G. Khlebtsov, Gold nanoparticles in biology and medicine: recent advances and prospects, Acta Naturae 3(2) (2011) 34–55.
65. X.-F. Zhang, Z.-G. Liu, W. Shen, S. Gurunathan, Silver Nanoparticles: Synthesis, Characterization, Properties, Applications, and Therapeutic Approaches, International journal of molecular sciences 17(9) (2016) 1534.
66. M.M. Kemp, A. Kumar, S. Mousa, E. Dyskin, M. Yalcin, P. Ajayan, R.J. Linhardt, S.A. Mousa, Gold and silver nanoparticles conjugated with heparin derivative possess anti-angiogenesis properties, Nanotechnology 20(45) (2009) 455104.
67. R.S. Darweesh, N.M. Ayoub, S. Nazzal, Gold nanoparticles and angiogenesis: molecular mechanisms and biomedical applications, Int J Nanomedicine 14 (2019) 7643–7663.

68. Z.F. Zhong, P.M. Hoi, G.S. Wu, Z.T. Xu, W. Tan, X.P. Chen, L. Cui, T. Wu, Y.T. Wang, Anti-angiogenic effect of furanodiene on HUVECs in vitro and on zebrafish in vivo, Journal of ethnopharmacology 141(2) (2012) 721–7.
69. F. Pan, W. Yang, W. Li, X.Y. Yang, S. Liu, X. Li, X. Zhao, H. Ding, L. Qin, Y. Pan, Conjugation of gold nanoparticles and recombinant human endostatin modulates vascular normalization via interruption of anterior gradient 2-mediated angiogenesis, Tumour biology : the journal of the International Society for Oncodevelopmental Biology and Medicine 39(7) (2017) 1010428317708547.
70. B.A. Saeed, V. Lim, N.A. Yusof, K.Z. Khor, H.S. Rahman, N. Abdul Samad, Antiangiogenic properties of nanoparticles: a systematic review, Int J Nanomedicine 14 (2019) 5135–5146.
71. D.C. Kennedy, B. Coen, A.M. Wheatley, K.J.A. McCullagh, Microvascular Experimentation in the Chick Chorioallantoic Membrane as a Model for Screening Angiogenic Agents including from Gene-Modified Cells, International journal of molecular sciences 23(1) (2021) 452.
72. S. Vimalraj, T. Ashokkumar, S. Saravanan, Biogenic gold nanoparticles synthesis mediated by Mangifera indica seed aqueous extracts exhibits antibacterial, anticancer and anti-angiogenic properties, Biomedicine & Pharmacotherapy 105 (2018) 440–448.
73. F. Pan, W. Li, W. Yang, X.Y. Yang, S. Liu, X. Li, X. Zhao, H. Ding, L. Qin, Y. Pan, Anterior gradient 2 as a supervisory marker for tumor vessel normalization induced by anti-angiogenic treatment, Oncol Lett 16(3) (2018) 3083–3091.
74. W. Li, X. Li, S. Liu, W. Yang, F. Pan, X.Y. Yang, B. Du, L. Qin, Y. Pan, Gold nanoparticles attenuate metastasis by tumor vasculature normalization and epithelial-mesenchymal transition inhibition, Int J Nanomedicine 12 (2017) 3509–3520.
75. P. Mukherjee, R. Bhattacharya, P. Wang, L. Wang, S. Basu, J.A. Nagy, A. Atala, D. Mukhopadhyay, S. Soker, Antiangiogenic Properties of Gold Nanoparticles, Clinical Cancer Research 11(9) (2005) 3530–3534.
76. S.K. Nune, P. Gunda, P.K. Thallapally, Y.-Y. Lin, M.L. Forrest, C.J. Berkland, Nanoparticles for biomedical imaging, Expert Opin Drug Deliv 6(11) (2009) 1175–1194.
77. M.S. Chavali, M.P. Nikolova, Metal oxide nanoparticles and their applications in nanotechnology, SN Applied Sciences 1(6) (2019) 607.
78. T. Naseem, T. Durrani, The role of some important metal oxide nanoparticles for wastewater and antibacterial applications: A review, Environmental Chemistry and Ecotoxicology 3 (2021) 59–75.
79. M. Divya, B. Vaseeharan, M. Abinaya, S. Vijayakumar, M. Govindarajan, N.S. Alharbi, S. Kadaikunnan, J.M. Khaled, G. Benelli, Biopolymer gelatin-coated zinc oxide nanoparticles showed high antibacterial, antibiofilm and anti-angiogenic activity, Journal of Photochemistry and Photobiology B: Biology 178 (2018) 211–218.
80. H. He, Z. Zou, B. Wang, G. Xu, C. Chen, X. Qin, C. Yu, J. Zhang, Copper Oxide Nanoparticles Induce Oxidative DNA Damage and Cell Death via Copper Ion-Mediated P38 MAPK Activation in Vascular Endothelial Cells, Int J Nanomedicine 15 (2020) 3291–3302.
81. D. Raj Preeth, M. Shairam, N. Suganya, R. Hootan, R. Kartik, K. Pierre, C. Suvro, S. Rajalakshmi, Green synthesis of copper oxide nanoparticles using sinapic acid: an underpinning step towards antiangiogenic therapy for breast cancer, Journal of biological inorganic chemistry : JBIC : a publication of the Society of Biological Inorganic Chemistry 24(5) (2019) 633–645.
82. C.J. Santos, D.C. Ferreira Soares, C.A. Ferreira, A.L.B. de Barros, A.D. Silva Cunha Junior, F.M. Filho, Antiangiogenic evaluation of ZnWO(4) nanoparticles synthesised through microwave-assisted hydrothermal method, Journal of drug targeting 26(9) (2018) 806–817.
83. K.E. Johnson, T.A. Wilgus, Vascular Endothelial Growth Factor and Angiogenesis in the Regulation of Cutaneous Wound Repair, Adv Wound Care (New Rochelle) 3(10) (2014) 647–661.

84. H. Song, W. Wang, P. Zhao, Z. Qi, S. Zhao, Cuprous oxide nanoparticles inhibit angiogenesis via down regulation of VEGFR2 expression, Nanoscale 6(6) (2014) 3206–3216.
85. M.S. Lord, B. Tsoi, C. Gunawan, W.Y. Teoh, R. Amal, J.M. Whitelock, Anti-angiogenic activity of heparin functionalised cerium oxide nanoparticles, Biomaterials 34(34) (2013) 8808–18.
86. H.O. Alsaab, A.S. Al-Hibs, R. Alzhrani, K.K. Alrabighi, A. Alqathama, A. Alwithenani, A.H. Almalki, Y.S. Althobaiti, Nanomaterials for Antiangiogenic Therapies for Cancer: A Promising Tool for Personalized Medicine, International journal of molecular sciences 22(4) (2021) 1631.
87. A.K. Barui, S.K. Nethi, S. Haque, P. Basuthakur, C.R. Patra, Recent Development of Metal Nanoparticles for Angiogenesis Study and Their Therapeutic Applications, ACS Applied Bio Materials 2(12) (2019) 5492–5511.
88. S. Mukherjee, Recent progress toward antiangiogenesis application of nanomedicine in cancer therapy, Future Science OA 4(7) (2018) FSO318.
89. S. Giri, A. Karakoti, R.P. Graham, J.L. Maguire, C.M. Reilly, S. Seal, R. Rattan, V. Shridhar, Nanoceria: A Rare-Earth Nanoparticle as a Novel Anti-Angiogenic Therapeutic Agent in Ovarian Cancer, PLOS ONE 8(1) (2013) e54578.
90. T. Yadavalli, D. Shukla, Role of metal and metal oxide nanoparticles as diagnostic and therapeutic tools for highly prevalent viral infections, Nanomedicine: Nanotechnology, Biology and Medicine 13(1) (2017) 219–230.
91. K.R. Singh, V. Nayak, J. Singh, A.K. Singh, R.P. Singh, Potentialities of bioinspired metal and metal oxide nanoparticles in biomedical sciences, RSC advances 11(40) (2021) 24722–24746.
92. K. Mylkie, P. Nowak, P. Rybczynski, M. Ziegler-Borowska, Polymer-Coated Magnetite Nanoparticles for Protein Immobilization, Materials 14(2) (2021) 248.
93. D. Banerjee, R. Harfouche, S. Sengupta, Nanotechnology-mediated targeting of tumor angiogenesis, Vasc Cell 3(1) (2011) 3–3.
94. H. Basoglu, B. Goncu, F. Akbas, Magnetic nanoparticle-mediated gene therapy to induce Fas apoptosis pathway in breast cancer, Cancer gene therapy 25(5-6) (2018) 141–147.
95. P.J.H. D.V.M, L.M.D.V.M. Karen, A.P. Alicia, D.P. James, S. Shawntel, J.G. David, F.S. Nicole, A.V. Frank, A.B. Alicea, J.W. Robert, R. Ashish, D. Danielle, C.-B. Margaret, N.F. Steven, Hypo-fractionated radiation, magnetic nanoparticle hyperthermia and a viral immunotherapy treatment of spontaneous canine cancer, Proc.SPIE, 2017.
96. Y.-X.J. Wang, Current status of superparamagnetic iron oxide contrast agents for liver magnetic resonance imaging, World J Gastroenterol 21(47) (2015) 13400–13402.
97. S. Natesan, C. Ponnusamy, A. Sugumaran, S. Chelladurai, S. Shanmugam Palaniappan, R. Palanichamy, Artemisinin loaded chitosan magnetic nanoparticles for the efficient targeting to the breast cancer, International journal of biological macromolecules 104(Pt B) (2017) 1853–1859.
98. A.K. Das, Anticancer Effect of AntiMalarial Artemisinin Compounds, Ann Med Health Sci Res 5(2) (2015) 93–102.
99. M.A. Abakumov, N.V. Nukolova, M. Sokolsky-Papkov, S.A. Shein, T.O. Sandalova, H.M. Vishwasrao, N.F. Grinenko, I.L. Gubsky, A.M. Abakumov, A.V. Kabanov, V.P. Chekhonin, VEGF-targeted magnetic nanoparticles for MRI visualization of brain tumor, Nanomedicine: Nanotechnology, Biology and Medicine 11(4) (2015) 825–833.
100. M.H. Karimi, G.R. Mahdavinia, B. Massoumi, pH-controlled sunitinib anticancer release from magnetic chitosan nanoparticles crosslinked with κ-carrageenan, Materials Science and Engineering: C 91 (2018) 705–714.
101. K.D. Son, Y.-J. Kim, Anticancer activity of drug-loaded calcium phosphate nanocomposites against human osteosarcoma, Biomaterials Research 21(1) (2017) 13.

102. Z. Sun, W. Li, J.C. Lenzo, J.A. Holden, M.J. McCullough, A.J. O'Connor, N.M. O'Brien-Simpson, The Potential of Calcium Phosphate Nanoparticles as Adjuvants and Vaccine Delivery Vehicles, Frontiers in Materials 8 (2021).
103. A. Maurya, A.K. Singh, G. Mishra, K. Kumari, A. Rai, B. Sharma, G.T. Kulkarni, R. Awasthi, Strategic use of nanotechnology in drug targeting and its consequences on human health: A focused review, Interv Med Appl Sci 11(1) (2019) 38–54.
104. K. Khosravi-Darani, M.R. Mozafari, L. Rashidi, M. Mohammadi, Calcium based non-viral gene delivery: an overview of methodology and applications, Acta medica Iranica 48(3) (2010) 133–41.
105. I. Roy, S. Mitra, A. Maitra, S. Mozumdar, Calcium phosphate nanoparticles as novel non-viral vectors for targeted gene delivery, International journal of pharmaceutics 250(1) (2003) 25–33.
106. K.-W. Huang, Y.-T. Lai, G.-J. Chern, S.-F. Huang, C.-L. Tsai, Y.-C. Sung, C.-C. Chiang, P.-B. Hwang, T.-L. Ho, R.-L. Huang, T.-Y. Shiue, Y. Chen, S.-K. Wang, Galactose Derivative-Modified Nanoparticles for Efficient siRNA Delivery to Hepatocellular Carcinoma, Biomacromolecules 19(6) (2018) 2330–2339.
107. N. Khatri, D. Baradia, I. Vhora, M. Rathi, A. Misra, Development and characterization of siRNA lipoplexes: Effect of different lipids, in vitro evaluation in cancerous cell lines and in vivo toxicity study, AAPS PharmSciTech 15(6) (2014) 1630–1643.
108. D. Maiti, X. Tong, X. Mou, K. Yang, Carbon-Based Nanomaterials for Biomedical Applications: A Recent Study, Frontiers in Pharmacology 9 (2019).
109. P. Sarkar, K. Ghosal, D. Chakraborty, K. Sarkar, Chapter 20 - Biocompatibility and biomedical applications of various carbon-based materials, in: S. Thomas, C. Sarathchandran, S.A. Ilangovan, J.C. Moreno-Piraján (Eds.), Handbook of Carbon-Based Nanomaterials, Elsevier2021, pp. 829–875.
110. Z. Liu, S. Tabakman, K. Welsher, H. Dai, Carbon Nanotubes in Biology and Medicine: In vitro and in vivo Detection, Imaging and Drug Delivery, Nano Res 2(2) (2009) 85–120.
111. S. Nasir, M.Z. Hussein, Z. Zainal, N.A. Yusof, Carbon-Based Nanomaterials/Allotropes: A Glimpse of Their Synthesis, Properties and Some Applications, Materials (Basel) 11(2) (2018) 295.
112. A.T. Smith, A.M. LaChance, S. Zeng, B. Liu, L. Sun, Synthesis, properties, and applications of graphene oxide/reduced graphene oxide and their nanocomposites, Nano Materials Science 1(1) (2019) 31–47.
113. K.V. Krishna, C. Ménard-Moyon, S. Verma, A. Bianco, Graphene-based nanomaterials for nanobiotechnology and biomedical applications, Nanomedicine 8(10) (2013) 1669–1688.
114. J. Liu, J. Dong, T. Zhang, Q. Peng, Graphene-based nanomaterials and their potentials in advanced drug delivery and cancer therapy, Journal of Controlled Release 286 (2018) 64–73.
115. Y. Tao, L. Zhu, Y. Zhao, X. Yi, L. Zhu, F. Ge, X. Mou, L. Chen, L. Sun, K. Yang, Nano-graphene oxide-manganese dioxide nanocomposites for overcoming tumor hypoxia and enhancing cancer radioisotope therapy, Nanoscale 10(11) (2018) 5114–5123.
116. X. Zhang, W. Tian, X. Cai, X. Wang, W. Dang, H. Tang, H. Cao, L. Wang, T. Chen, Hydrazinocurcumin Encapsuled nanoparticles "re-educate" tumor-associated macrophages and exhibit anti-tumor effects on breast cancer following STAT3 suppression, PLoS One 8(6) (2013) e65896.

117. N. Malhotra, G. Audira, A.L. Castillo, P. Siregar, J.M.S. Ruallo, M.J. Roldan, J.-R. Chen, J.-S. Lee, T.-R. Ger, C.-D. Hsiao, An Update Report on the Biosafety and Potential Toxicity of Fullerene-Based Nanomaterials toward Aquatic Animals, Oxid Med Cell Longev 2021 (2021) 7995223–7995223.
118. M. Fiorillo, A.F. Verre, M. Iliut, M. Peiris-Pagés, B. Ozsvari, R. Gandara, A.R. Cappello, F. Sotgia, A. Vijayaraghavan, M.P. Lisanti, Graphene oxide selectively targets cancer stem cells, across multiple tumor types: implications for non-toxic cancer treatment, via "differentiation-based nano-therapy", Oncotarget 6(6) (2015) 3553–62.
119. Z. Chen, R. Mao, Y. Liu, Fullerenes for cancer diagnosis and therapy: preparation, biological and clinical perspectives, Current drug metabolism 13(8) (2012) 1035–45.
120. P. Mroz, G.P. Tegos, H. Gali, T. Wharton, T. Sarna, M.R. Hamblin, Photodynamic therapy with fullerenes, Photochem Photobiol Sci 6(11) (2007) 1139–1149.
121. T. Maruyama, Chapter 6-Carbon nanotubes, in: S. Thomas, C. Sarathchandran, S.A. Ilangovan, J.C. Moreno-Piraján (Eds.), Handbook of Carbon-Based Nanomaterials, Elsevier2021, pp. 299–319.
122. S.Y. Madani, N. Naderi, O. Dissanayake, A. Tan, A.M. Seifalian, A new era of cancer treatment: carbon nanotubes as drug delivery tools, Int J Nanomedicine 6 (2011) 2963–2979.
123. M. Das, P. Nariya, A. Joshi, A. Vohra, R. Devkar, S. Seshadri, S. Thakore, Carbon nanotube embedded cyclodextrin polymer derived injectable nanocarrier: A multiple faceted platform for stimulation of multi-drug resistance reversal, Carbohydrate Polymers 247 (2020) 116751.
124. M. Rajabi, S.A. Mousa, The Role of Angiogenesis in Cancer Treatment, Biomedicines 5(2) (2017) 34.
125. S. Murugesan, S.A. Mousa, J. O'Connor L, D.W. Lincoln, 2nd, R.J. Linhardt, Carbon inhibits vascular endothelial growth factor- and fibroblast growth factor-promoted angiogenesis, FEBS letters 581(6) (2007) 1157–60.
126. M.M. Barroso, Quantum dots in cell biology, J Histochem Cytochem 59(3) (2011) 237–251.
127. M. d'Amora, S. Giordani, 7 - Carbon Nanomaterials for Nanomedicine, in: G. Ciofani (Ed.), Smart Nanoparticles for Biomedicine, Elsevier2018, pp. 103–113.
128. C.S.M. Martins, A.P. LaGrow, J.A.V. Prior, Quantum Dots for Cancer-Related miRNA Monitoring, ACS Sensors (2022).
129. D. Chen, B. Li, T. Lei, D. Na, M. Nie, Y. Yang, Congjia, Xie, Z. He, J. Wang, Selective mediation of ovarian cancer SKOV3 cells death by pristine carbon quantum dots/Cu2O composite through targeting matrix metalloproteinases, angiogenic cytokines and cytoskeleton, J Nanobiotechnology 19(1) (2021) 68.

Potential Challenges in Anti-angiogenic Nanotherapies and Conclusion

6

6.1 Introduction

Since last few decades, several nanoparticles-based medicines were tested and clinically used for various biomedical applications including cancer [1, 2]. Unfortunately, many nanoparticles-based materials demonstrate short-term or long-term toxicity in human or environmental species. Hence, it is critical to collect detailed information on various aspects of these nanomaterials including acute toxicity, chronic toxicity, pharmacokinetics, excretion, metabolic pathways, immunogenicity, pharmacodynamics, eco-toxicity etc. in animal models prior to applying the these in human clinical trials. The potential challenge is to develop effective and robust anti-angiogenic nanomedicine, which will explicitly target the malignant cells, without displaying notable toxicity in healthy cells, tissues, and organs. Additionally, existing FDA approved anti-angiogenic drugs to treat cancer can be conjugated with these novel nanomaterials to combinatorially used as an improved therapeutic approach. However, various issues need to be solved in detail before a product can be clinically translated: (a) challenges in industrial scale productions, (b) challenges related to diffusion, (c) challenges regarding acute and chronic toxicity of the nanoparticles, (d) challenges regarding uptake and active targeting of the nanomedicine, (e) challenges to identify the best route of administration and dosing regimen and (f) challenges to find the biodegradable formulations etc. Academic researchers need to closely work with medical, industrial and regulatory partners to create a well-documented strategy that can be easily followed by many to translate some of these exciting nanomedical technologies to treat various deadly diseases including cancers [3, 4].

S. Mukherjee and V. S. Madamsetty, *Nanoparticles in Angiogenesis and Cancer*,
Synthesis Lectures on Biomedical Engineering,
https://doi.org/10.1007/978-3-031-11284-3_6

6.1.1 Challenges and Difficulties of Production

In general, industrial scale productions of nanomaterials have various obstacles and challenges during scale up process including reproducibility, cost, purification, aggregation, well dispersibility etc. [5]. It is needless to say that the industrial scale up or scale down process are much complex than from laboratory-based methods. Commercial manufacturing requires careful control of several parameters including temperature, volume, pH, mixing conditions etc. [5]. that can cause several layers of production challenges. Additionally, use of high temperature reactors, emulsification, milling, sonication, high-speed homogenization, centrifugation, filtration, or lyophilization can significantly change the difficulty levels during industrial scale productions [5]. It will be great if the researchers can consider a robust synthesis approach during the initial lab scale development, which will facilitate better chance for upscale industrial manufacturing. In addition, it is extremely important to recognize and note all the synthesis requirements including temperature, pressure, the concentration of reactants, molar ratio of reactants, the reagent mixing conditions, reaction time, reagents adding rate, stirring conditions, solvent types, stabilizer/crosslinker details, water-to oil phase ratio and reaction pH etc. that may play a role to change the synthesis yield, characteristics, and usefulness of the synthesized nanomaterials. Additionally, it is critical to calculate the number of targeting agents or active drugs to be attached to each nanoparticle as slight alteration of these conditions can cause unstable and unsought products that are biologically ineffective along with a great quantity of impurity. Biosynthesized nanoparticles products made by plant, fungus, algae etc. have additional challenges with seasonal changes that impact the quality and bioactivity of the final material. Hence, the laboratory scale product development stages need to be robust with superior reproducibility, properly characterized and analyzed for the biological activities in order to have a better chance of upscale industrial manufacturing for commercial use.

6.1.2 Targeting

Generally, nanoparticles and nanoconjugates can be utilized in two ways to target the disease site in living body. Firstly, attaching any active target agent to the surface of nanomaterials that will trigger an active targeting approach using specific receptor binding. Secondly, nanomaterials can passively target to the disease site especially by fast pass metabolism and enhanced permeability and retention (EPR) effect in case of leaky tumor vasculature [6]. Nanomaterials can easily penetrate the leaky tumor vasculature, but due to lack of efficient lymphatic drainage they get retained inside the tumor vasculature causing the EPR effects and passive targeting [6]. However, passive targeting approaches are not effective, causing non-specific toxicity in healthy cells/tissues or organs [7]. To overcome these challenges several approaches of highly specific active targeting agents are being used that may help to increase the nanoconjugate circulation, tumor selective uptake and bioavailability in tumor microenvironment. This will help also to obtain improved therapeutic efficacy along with lower toxicity of anti-angiogenic agents. Several active targeting agents are now being tested for anti-angiogenic based cancer therapy including monoclonal antibodies, αvβ3 integrin receptor antibody, antiangiogenic peptides (eg, RGD/NGR) etc. [8]. Integrins are a class of family of cell surface receptors (αvβ3 or αvβ5) that has observed to be radically upregulated in the process of neovascularization during cancer progression, migration, and metastasis [8]. Utilization of active targeting agents, to target angiogenic signaling using αvβ3 integrin antibody or anti-angiogenic peptides has dramatically improved the therapeutic efficiency of the nanoparticle drug conjugates and reduced the non-specific toxicity to the healthy organs. However, recent reports state that less than 1% of inserted nanomaterials is transported to the disease sites either for active or passive targeting approaches. Hence, more research is required to facilitate the biodistribution and bioavailability of the anti-angiogenic nanomedicine to treat tumor malignancy.

6.1.3 The Diffusion

It is important to discuss that diffusion and penetration of nanomaterials to cell and tissue that present a critical test for enhanced uptake and improved effectiveness of the anti-angiogenic nanomedicine. Nanoparticle-based medicine face several biological obstacles, including extracellular matrix (ECM) and mucus gels, which needs to bypass before they can achieve entry to the target sites [9]. ECM has a mesh-like architecture with a cross-linking and dense network of collagen fibers inhibiting effective particle transport that causes reduction in the delivery of the nanomedicine to the target disease sites. During cancer, the ECM develop into a denser microenvironment causing further challenges for effective penetration of nanomedicines. Many different factors play a major role for effective intravascular delivery of nanoparticles, including (i) early immune rejection and macrophage uptake, (ii) fast pass metabolism, (iii) clearance through the liver and spleen, (iv) infusion across the cell-tissue endothelium into and the interstitial tissue and target disease sites, (vi) endocytosis or receptor arbitrated entrance in target sites, and (vii) cytoplasmic/nucleus/lysosomal diffusion-based entry and release of anti-angiogenic agents [10]. Additional option of delivery routes through skin, nasal, vagina, muscle membranes possess severe to carry the nanomedicine through the interstitial tissue by avoiding the diffusive resistance. Size, shape, density, surface chemistry, peptide coatings, and sedimentation velocities can play crucial role for improved diffusion of nanomaterials and can reduce problems associated with diffusion-related.

6.1.4 Toxicity and Immunogenicity

Toxicity and immunogenicity of the injected materials including anti-angiogenic nanomaterials are important factors to consider for successful clinical therapeutic applications [11–14]. Insight of risk associate with human health with exposure of new materials including nanomaterials is necessary before these technologies can be clinically translated from laboratory scale research. Imminent toxicity associated with different nanomaterials cause an immense challenge for healthcare applications when being used at various routes of exposure. Many reports demonstrated acute and chronic toxicity symptoms of different nanoparticles including silver, zinc oxide, copper oxide, platinum, cobalt oxide, cerium oxide, quantum dots etc. [15–17]. However, various groups also showed non-toxic character of different synthesized nanomaterials including gold nanoparticles, platinum nanoparticles, silver Prussian blue nanoparticles, europium hydroxide nanorods: EHN, zinc oxide nanoflowers: ZONF and silica-based nanoparticles etc. [15–17]. towards in vivo mouse and rat models. Additionally, several other parameters related to nanoparticles pharmacokinetics, bio-distribution and clearance were studied in vivo. Mukherjee et al. recently demonstrated in vivo biocompatibility of pegylated platinum nanoparticles upto 10 $mgkg^{-1}$ dose in C57BL6/J mouse administered with a single intraperitoneal dose (IP) using histological analysis and blood-serum parameter analysis (Fig. 6.1) [18]. Whereas, at a higher dose of 50 $mgkg^{-1}$ dose it displayed slight nephrotoxicity. These nanomaterials were further utilized as a drug delivery vehicle to deliver anticancer drugs in melanoma tumor model, successfully reducing the tumor burden [19]. In another report, Mukherjee and co-workers, demonstrated the in vivo biocompatibility of polyvinylpyrrolidone coated silver Prussian blue analogue nanoparticles that were used for multifunctional biomedical applications in cancer therapy and anti-bacterial therapy [20]. Hence, these strategies demonstrate the use of biocompatible polymer coating to reduce or eliminate any toxicity of the nanomaterials that can be used for clinical applications. Other reports analyzed the size, shape and morphological effects of nanomaterials to their associated toxicity [21].

Das et al. demonstrated the synthesis of dumbbell shaped vanadium pentoxide nanoparticles (V_2O_5 NPs) with a size of 30–60 nm that showed no toxicity in vivo and utilized for anti-angiogenic cancer therapy in C57BL6/J melanoma tumor model [22]. Liu et al. showed low toxicity (based on overall mortality, histopathological studies, hematological parameters) of mesoporous silica nanoparticles following single intravenous (IV) injection or repeated dosing up to 80 $mgkg^{-1}$ in mouse models. Gold, ZONF, EHNs were found as nontoxic towards in vivo mouse model tested for acute and chronic studies [23]. However, contrary to these works, there are also published reports of acute and chronic toxicity, nephrotoxicity, and liver toxicity of several nanomaterials including cerium oxide, zinc nanoparticles, platinum nanoparticles etc. [24–28]. Hence, it becomes absolute necessary to carefully evaluate the toxicity and other immunological responses of the nanomaterials before administrating to the human. A current report by Hansen and coworkers reviewed about 400 findings with 965 nanoparticles to address the potential mammalian toxicity,

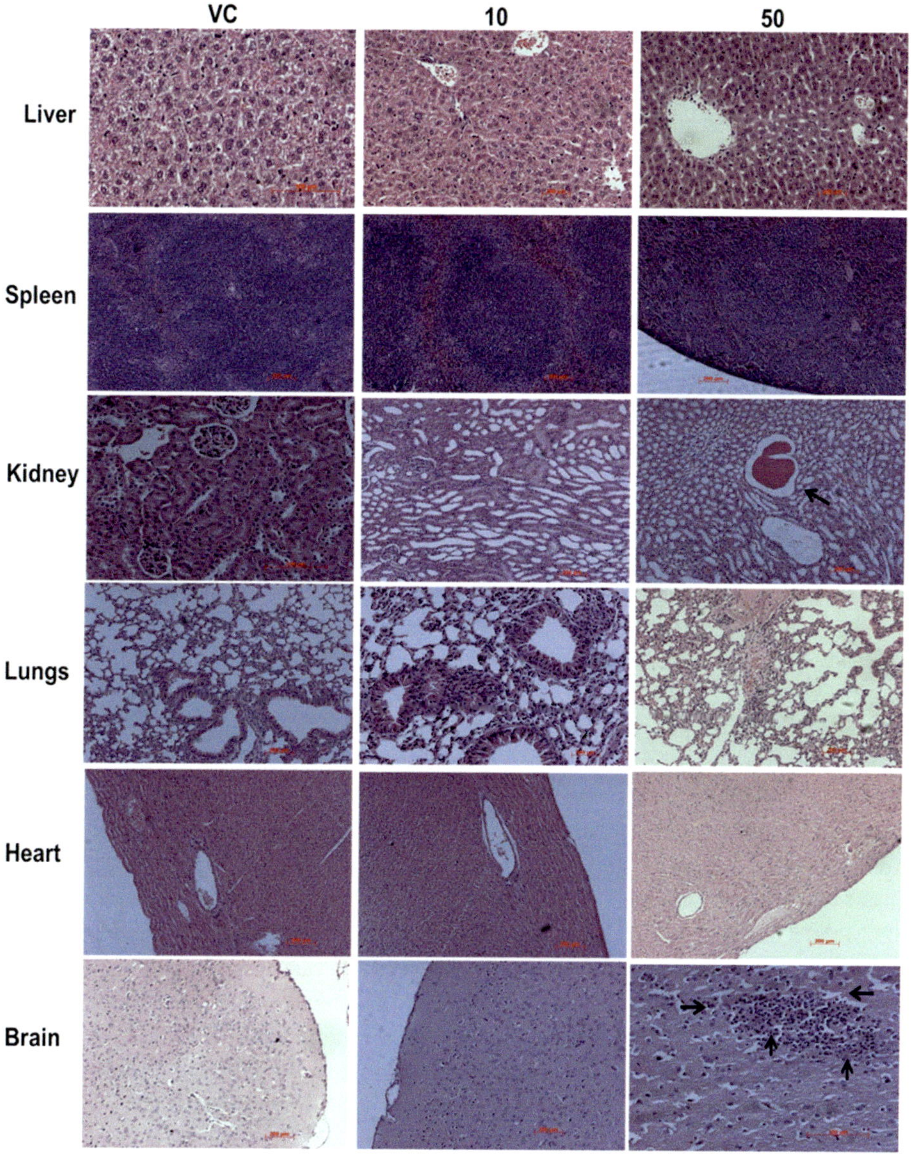

Fig. 6.1 Histopathological study of essential organs. Sections of the different organs of control and treated C57BL6/J female mice treated with PEG6000-PtNPs (containing PEG6000) (10 and 50 mg kg− 1 b.w.) were analyzed using H&E staining. Five mice for each group were studied. Figure is reproduced with permission from Wiley publishers [18]

and environmental toxicity of these various materials [29]. Additionally, immunogenicity and antigenicity of nanomaterials are being investigated to make substantial progress in understanding the reason that makes a nanoparticle immunogenic, interactions of immune cells to nanoparticles, and their eventual challenges for drug delivery applications [30].

6.1.5 Biodegradability and Clearance

Biodegradability and facile excretion of the nanomaterials are very critical to avoid long term accumulation and toxicity in the body. Long term accumulation of nanomaterials can severely impact the usefulness of the nanomaterials and reduce the clinical significance. It is well known that organic nanomaterials including liposomes, polymers, micelles, lipid nanoparticles, dendrimers etc. demonstrate excellent degradability with time that has been excreted from body within rapid time. Though, many of the current metal-based nanoparticles have issues with slow excretion metabolic degradation, causing severe challenges and exhibiting toxicity. Only some current findings revealed that metal-based nanomaterials are gradually excreted through hepatobiliary routes by urine and feces. Still, detailed mechanistic studies are not well carried out on metabolic and excretory pathways that limited the knowledge and clinical significance. In a recent report Mukherjee et al. performed a toxicity and clearance study in vivo in C57BL6/J mouse model after administration of pegylated platinum nanoparticles (PEG-PtNPs) in different doses (10 and 50 mg kg^{-1}) injected intraperitoneally [18]. Data showed that PEG-PtNPs were mostly distributed in spleen followed by in liver and tail confirmed by ICP-OES analysis [18]. It is predicted that these patterns of biodistribution is mostly regulated by the reticuloendothelial system (RES) of body, which is an important component of the human immune structure. Additionally, pharmacokinetics studies were carried out using a non-compartmental PK model that showed the time (T_{max}) needed for achieving highest concentration (C_{max}) of PEG-PtNPs in blood plasma is 1 h after the injection of nanoparticles. Moreover, the time required to reduce 50% of PEG-PtNPs in the blood plasma was obtained to be $\approx$135 h indicating prolonged removal from the blood [18]. The feces and urine were collected that showed the presence of platinum nanoparticles confirmed by ICP-OES analysis, suggesting the possible excretion through hepatobiliary systems [18]. In another recent report, Das et al. demonstrated the excretion studies of anti-angiogenic dumbbell shaped vanadium pentoxide nanoparticles (V_2O_5 NPs), indicating slow release of vanadium in both urine and feces even after four weeks of NPs administration in mouse model [22]. These results show the nanoparticles clearance through hepatobiliary route. However, detailed studies are required to identify the exact mechanisms of excretion of these nanoparticles.

Cassano and co-workers reported complete biodegradation of gold nanoparticles encapsulated in silica nanospheres (AuSi) through renal clearance, that avoids unwanted tissue accumulation and related toxicity [31]. It was predicted that silica nanospheres may turn soluble into silicic acid and further excreted by urine. Leftover AuNPs may get covered with the endogenic glutathione that can facilities the renal excretion. Choi et al. showed the facile clearance of 5.5 nm sized quantum dots nanoparticles from the urinary route in a rodent model [32]. Rengan and coworkers reported metabolic degradation of liposomal gold nanoparticles (Lipos-AuNPs) in the liver and hepatocytes and facile elimination by hepatobiliary route up to 14 days [33]. ICP-MS analysis confirmed gradual clearance of the nanoparticles from blood plasma. Moreover, authors predicted that positively charged AuNPs might prevail over charge repulsion from negative charged glomerular basement membrane that is present in the nephrons to cleared through renal routes. These various reports confirm that size, shape, surface morphology, coating, surface charge, concentration, viscosity, and density can play important roles to modulate the absorption, plasma kinetics, transport, pharmacokinetics, bio-distribution, accumulation, excretion, and consequent biomolecular properties of the administered nanomaterials in any living systems. Extensive studies are required to further understand the challenges of degradation and necessary steps need to be taken before clinical applications of anti-angiogenic nanomedicine.

6.2 Conclusions

The development of nanobiotechnologies with fast expanding industry could look ahead to the potential answer for cancer therapy and diagnosis following significant actions taken regarding several safety aspects and detailed mechanistic investigations. At present several research groups are working on the investigation of new anti-angiogenic nanomedicines for cancer therapy to prevent tumor growth, metastasis and improving life expectancy of cancer patients. These nanomedicines alone or in combination with the existing treatment options can work as a combinatorial mode for complete eradication of this deadly disease. Due to drug resistance and unavoidable side effects many anti-angiogenic drugs facilitated cancer treatments are restricted for the clinical uses. Since past decade, nanobiotechnology based novel approaches has been employed for multifunctional actions in healthcare. These recent developments of the anti-angiogenic nanomaterials will transform the challenges of the existing strategies associated with the conventional medicines. Once we have a detailed pictures on the potential toxicity of these nanoparticles and potential impacts of these in human health and environment, regulatory agencies will feel more comfortable to clinically approve these technologies for the use of human. Additionally, these anti-angiogenic nanomaterials are either being provided alone or in combination with other active therapeutic agents to facilitate the rapid cure. However, significant attentions need to be provided to determine long term toxicity, metabolic long-term fate pharmacokinetics, biodegradation, immunogenicity, pharmacodynamics, other safety issues and production challenges before these new technologies get commercially useful. This present book highlights various aspects of the new generation anti-angiogenic nanomaterials and their possible uses in different cancer therapies. The nanomedicine-based methodology permits researchers to create innovative targeted anti-angiogenic nanomaterials, which might be extremely encouraging and achievable replacement tools to cure cancer soon.

References

1. S. Mukherjee, D. Sonanini, A. Maurer, H.E. Daldrup-Link, The yin and yang of imaging tumor associated macrophages with PET and MRI, Theranostics 9(25) (2019) 7730-7748.
2. S. Mukherjee, M. V., D. Bhattacharya, S.R. Chowdhury, M.K. Paul, A. Mukherjee, Recent Advancements of Nanomedicine in Neurodegenerative Disorders Theranostics, Advanced Functional Materials 30(35) 2003054.
3. A. Mukherjee, V.S. Madamsetty, M.K. Paul, S. Mukherjee, Recent Advancements of Nanomedicine towards Antiangiogenic Therapy in Cancer, Int J Mol Sci 21(2) (2020).
4. A. Mukherjee, M. Paul, S. Mukherjee, Recent Progress in the Theranostics Application of Nanomedicine in Lung Cancer, Cancers (Basel) 11(5) (2019).
5. N. Desai, Challenges in development of nanoparticle-based therapeutics, AAPS J 14(2) (2012) 282–95.
6. R. Ngoune, A. Peters, D. von Elverfeldt, K. Winkler, G. Putz, Accumulating nanoparticles by EPR: A route of no return, J Control Release 238 (2016) 58–70.
7. J.J. Milligan, S. Saha, A Nanoparticle's Journey to the Tumor: Strategies to Overcome First-Pass Metabolism and Their Limitations, Cancers (Basel) 14(7) (2022).
8. R. Pasqualini, E. Koivunen, E. Ruoslahti, Alpha v integrins as receptors for tumor targeting by circulating ligands, Nat Biotechnol 15(6) (1997) 542–6.
9. S. Barua, S. Mitragotri, Challenges associated with Penetration of Nanoparticles across Cell and Tissue Barriers: A Review of Current Status and Future Prospects, Nano Today 9(2) (2014) 223–243.
10. E.C. Cho, Q. Zhang, Y. Xia, The effect of sedimentation and diffusion on cellular uptake of gold nanoparticles, Nat Nanotechnol 6(6) (2011) 385–91.
11. A.T. Iacob, F.G. Lupascu, M. Apotrosoaei, I.M. Vasincu, R.G. Tauser, D. Lupascu, S.E. Giusca, I.D. Caruntu, L. Profire, Recent Biomedical Approaches for Chitosan Based Materials as Drug Delivery Nanocarriers, Pharmaceutics 13(4) (2021).
12. D.H. Lee, S.Y. Choi, K.K. Jung, J.Y. Yang, J.Y. Jeong, J.H. Oh, S.H. Kim, J.H. Lee, The Research of Toxicity and Sensitization Potential of PEGylated Silver and Gold Nanomaterials, Toxics 9(12) (2021).
13. W. Liang, Y. Dong, R. Shao, S. Zhang, X. Wu, X. Huang, B. Sun, B. Zeng, J. Zhao, Application of nanoparticles in drug delivery for the treatment of osteosarcoma: focussing on the liposomes, J Drug Target 30(5) (2022) 463–475.
14. R. Solanki, H. Rostamabadi, S. Patel, S.M. Jafari, Anticancer nano-delivery systems based on bovine serum albumin nanoparticles: A critical review, Int J Biol Macromol 193(Pt A) (2021) 528–540.
15. S. Bayda, M. Hadla, S. Palazzolo, P. Riello, G. Corona, G. Toffoli, F. Rizzolio, Inorganic Nanoparticles for Cancer Therapy: A Transition from Lab to Clinic, Curr Med Chem 25(34) (2018) 4269–4303.
16. A. Bolandparvaz, N. Vapniarsky, R. Harriman, K. Alvarez, J. Saini, Z. Zang, J. Van De Water, J.S. Lewis, Biodistribution and toxicity of epitope-functionalized dextran iron oxide nanoparticles in a pregnant murine model, J Biomed Mater Res A 108(5) (2020) 1186–1202.

17. H.W. Despres, A. Sabra, P. Anderson, U.D. Hemraz, Y. Boluk, R. Sunasee, K. Ckless, Mechanisms of the immune response cause by cationic and anionic surface functionalized cellulose nanocrystals using cell-based assays, Toxicol In Vitro 55 (2019) 124–133.
18. S. Mukherjee, V.S. Bollu, A. Roy, S.K. Nethi, K. Madhusudana, J.M. Kumar, R. Sistla, C.R. Patra, Acute Toxicity, Biodistribution, and Pharmacokinetics Studies of Pegylated Platinum Nanoparticles in Mouse Model, Advanced NanoBiomed Research 1(7) (2021) 2000082.
19. S. Mukherjee, R. Kotcherlakota, S. Haque, D. Bhattacharya, J.M. Kumar, S. Chakravarty, C.R. Patra, Improved delivery of doxorubicin using rationally designed PEGylated platinum nanoparticles for the treatment of melanoma, Mater Sci Eng C Mater Biol Appl 108 (2020) 110375.
20. S. Mukherjee, R. Kotcherlakota, S. Haque, S. Das, S. Nuthi, D. Bhattacharya, K. Madhusudana, S. Chakravarty, R. Sistla, C.R. Patra, Silver Prussian Blue Analogue Nanoparticles: Rationally Designed Advanced Nanomedicine for Multifunctional Biomedical Applications, ACS Biomater Sci Eng 6(1) (2020) 690–704.
21. C. Egbuna, V.K. Parmar, J. Jeevanandam, S.M. Ezzat, K.C. Patrick-Iwuanyanwu, C.O. Adetunji, J. Khan, E.N. Onyeike, C.Z. Uche, M. Akram, M.S. Ibrahim, N.M. El Mahdy, C.G. Awuchi, K. Saravanan, H. Tijjani, U.E. Odoh, M. Messaoudi, J.C. Ifemeje, M.C. Olisah, N.J. Ezeofor, C.J. Chikwendu, C.G. Ibeabuchi, Toxicity of Nanoparticles in Biomedical Application: Nanotoxicology, J Toxicol 2021 (2021) 9954443.
22. S. Das, A. Roy, A.K. Barui, M.M.A. Alabbasi, M. Kuncha, R. Sistla, B. Sreedhar, C.R. Patra, Anti-angiogenic vanadium pentoxide nanoparticles for the treatment of melanoma and their in vivo toxicity study, Nanoscale 12(14) (2020) 7604–7621.
23. T. Liu, L. Li, X. Teng, X. Huang, H. Liu, D. Chen, J. Ren, J. He, F. Tang, Single and repeated dose toxicity of mesoporous hollow silica nanoparticles in intravenously exposed mice, Biomaterials 32(6) (2011) 1657–68.
24. F.S. Aljohani, M.T. Hamed, B.A. Bakr, Y.H. Shahin, M.M. Abu-Serie, A.K. Awaad, H. El-Kady, B.H. Elwakil, In vivo bio-distribution and acute toxicity evaluation of greenly synthesized ultra-small gold nanoparticles with different biological activities, Sci Rep 12(1) (2022) 6269.
25. P.E. Feuser, M.M. Cardoso, N.C. Galvani, R.P. Zaccaron, L.M. Venturini, F.K. Rigo, R.A. Machado-de-Avila, P.C.L. Silveira, C. Sayer, P.H. Hermes de Araujo, Evaluation of the in vivo acute toxicity of poly(thioether-ester) and superparamagnetic poly(thioether-ester) nanoparticles obtained by thiol-ene miniemulsion polymerization, J Biomed Mater Res B Appl Biomater 110(3) (2022) 702–711.
26. A. Shah, I. Tauseef, M.A. Yameen, S.K. Haleem, S. Haq, S. Shoukat, In-vivo toxicity and therapeutic efficacy of Paeonia emodi-mediated zinc oxide nanoparticles: In-vitro study, Microsc Res Tech 85(1) (2022) 181–192.
27. Q. Wang, Y. Zhu, B. Song, R. Fu, Y. Zhou, The In Vivo Toxicity Assessments of Water-Dispersed Fluorescent Silicon Nanoparticles in Caenorhabditis elegans, Int J Environ Res Public Health 19(7) (2022).
28. Z. Zhang, L. Zhao, Y. Ma, J. Liu, Y. Huang, X. Fu, S. Peng, X. Wang, Y. Yang, X. Zhang, W. Ding, J. Yu, Y. Zhu, H. Yan, S. Yang, Mechanistic study of silica nanoparticles on the size-dependent retinal toxicity in vitro and in vivo, J Nanobiotechnology 20(1) (2022) 146.
29. S. Foss Hansen, B.H. Larsen, S.I. Olsen, A. Baun, Categorization framework to aid hazard identification of nanomaterials, Nanotoxicology 1(3) (2007) 243–250.
30. A.N. Ilinskaya, M.A. Dobrovolskaia, Understanding the immunogenicity and antigenicity of nanomaterials: Past, present and future, Toxicol Appl Pharmacol 299 (2016) 70–7.
31. D. Cassano, D. Rota Martir, G. Signore, V. Piazza, V. Voliani, Biodegradable hollow silica nanospheres containing gold nanoparticle arrays, Chem Commun (Camb) 51(49) (2015) 9939–41.

32. H.S. Choi, W. Liu, P. Misra, E. Tanaka, J.P. Zimmer, B. Itty Ipe, M.G. Bawendi, J.V. Frangioni, Renal clearance of quantum dots, Nat Biotechnol 25(10) (2007) 1165–70.
33. A.K. Rengan, A.B. Bukhari, A. Pradhan, R. Malhotra, R. Banerjee, R. Srivastava, A. De, In vivo analysis of biodegradable liposome gold nanoparticles as efficient agents for photothermal therapy of cancer, Nano Lett 15(2) (2015) 842–8.

MIX
Papier aus verantwortungsvollen Quellen
Paper from responsible sources
FSC® C105338

If you have any concerns about our products,
you can contact us on
ProductSafety@springernature.com

In case Publisher is established outside the EU,
the EU authorized representative is:
Springer Nature Customer Service Center GmbH
Europaplatz 3, 69115 Heidelberg, Germany

Printed by Libri Plureos GmbH
in Hamburg, Germany